Partly to Mostly Funny

The Ultimate Weather Joke Book

EDITED BY
Jon Malay

WITH JOKES FROM
Norm Dvoskin

AMERICAN METEOROLOGICAL SOCIETY

Library of Congress Cataloging-in-Publication data
is available and can be found online at www.ametsoc.org.

I've lived in good climate, and it bores the
hell out of me. I like weather rather than climate.
—*John Steinbeck*

A lot of people like snow. I find it
to be an unnecessary freezing of water.
—*Carl Reiner*

The sun did not shine. It was too wet to play.
So we sat in the house. All that cold, cold, wet day.
—*Dr. Seuss, The Cat in the Hat*

Weather tonight: dark.
Turning partly light by morning.
—*George Carlin (as Al Sleet, the Hippy
Dippy Weatherman)*

Contents

Editor's Introduction

Thanks for buying or maybe borrowing this book from a friend. It means you're probably interested in weather, and you probably have a pretty good sense of humor. I undertook this project as a volunteer for the American Meteorological Society (AMS) for both those reasons, plus another, even better reason: The AMS is a great scientific and professional organization that plans to use the proceeds of sales of this book to support its important work. I invite you to learn more about the Society at www. ametsoc.org. If you're a professional in the atmospheric or related sciences, you should by all means be a Professional Member of the AMS. If you're not a pro but love weather and the oceans, or if you're interested in the concrete science behind climate change, please consider joining the

AMS as an Associate Member. Instructions on how to join are on the website. You won't regret it!

Now, before you go looking for laughs in this book, please consider where I'm coming from as its editor, and forgive me if this sounds like the setup of a joke. Maybe it is. I have three grown-up kids: a blonde, a lawyer, and an engineer. Actually, to be more precise, the lawyer is also a blonde! All three are very smart and successful young professionals, and each has a wonderful sense of humor. And yet, I'm not sure I can remember any of them telling blonde jokes, lawyer jokes, or engineer jokes despite there being no shortage of great material in all three categories of humor. That's a shame, because I heard a good one: "*A blonde, a lawyer, and an engineer walk into a bar. . . .*" But I digress.

I'm a retired Navy officer who's now a senior business development guy in the aerospace industry in Washington, D.C. But my academic degrees are in oceanography and meteorology, and I've been a professional meteorologist at various stages of my 40-year career. I recently served as president of the AMS, and needless to say, I'm very proud to have had the privilege of being elected by our members to that leadership position. As such, I'm capable of being a pretty serious guy when I need to, but the truth is I love hearing and telling jokes, and I've been telling jokes about the weather throughout my careers in the Navy and in the sciences industry. In fact, it's a tradition among my fellow past presidents to break into joke-telling at our annual dinner . . . and most of them are pretty good weather jokes. You might say telling weather jokes is in our DNA.

The AMS publishes highly respected scientific and technical journals, scientific monographs, and textbooks in the fields of atmospheric, oceanographic, and climate science. This begs the logical question, why a joke book? We're doing this book for the sheer fun of it. Well, OK, as I said earlier, we do hope to sell a ton of them so that we might raise funds for our activities within science education and public policy. So, true to its title, this is a book of *partly to mostly funny* weather jokes, and my colleagues at the AMS and I hope you enjoy it. I'll confess the title was chosen not just for the weather forecasting metaphor but because some of the jokes are really funny and some are genuinely lame. Unapologetically, we can assert pretty much all the classics are in here. And I decided to go for inclusiveness, only rejecting jokes that are either totally tasteless or hopelessly humorless.

I hope my colleagues in the weather business will also appreciate this treasury of jokes for professional reasons because, unlike that *other* category of people who jokes are written *about* and who get paid even when they're wrong (i.e., economists!), we meteorologists are in a profession that actually requires a sense of humor. I, for one, have presented bleak forecasts of bad weather to Navy admirals and commanding officers, many of whom were pretty much humorless individuals who didn't enjoy getting bad news, and they often felt obligated to share their innermost feelings about my professional abilities. Remember . . . a perfectly accurate forecast of bad weather is still a "bad forecast," right? And, since I have this aforementioned sense of humor, in those very stressful moments what I really wanted to do was to tell them a good weather joke. You know, to lighten the mood just a bit. Perhaps that was the seed of this book you now have at your fingertips. C'mon . . . let the sunshine of laughter brighten any (physically or metaphorically) cloudy day. As the song goes: "Gray skies are gonna clear up. Put on a happy face!"

This book wouldn't have been possible without the generous contribution of the vast majority of these jokes by AMS professional member Norm Dvoskin. Norm has had a long and successful career as an on-camera broadcast meteorologist on Long Island, New York. We solicited the submission of jokes from our members and collected a few pretty good ones, but when Norm contacted me to say he's been collecting weather jokes throughout his career, and offered to open up his treasure chest for use in this book, I knew we'd be able to pull this off. Nearly all of Norm's collection of groaners, one-liners, puns, and just funny stuff are between these covers and, unless we've identified another source with an annotation, we can thank Norm for his very generous contributions.

I also want to acknowledge the greatly appreciated inclusion of some very funny weather-themed cartoons from my fellow retired Navy meteorologist/oceanographer Jeff

Bacon. Jeff is a successful professional cartoonist whose hilarious "Broadside" cartoons have been appearing in the *Navy Times* publication and his own books for many years. His cartoons all come from his military background, but you'll appreciate his great sense of humor and ability to capture some of the more absurd moments of our business.

I've elected to group the jokes in roughly the same categories Norm had them organized by themes, such as the seasons, dry weather, or forecasting. This will enable you to pull up just the right joke for any occasion, even if there's a hurricane or a tornado bearing down on you. You might find yourself in a position to cheer your fellow weather-dodgers down there in the storm shelter!

And so I recommend you keep this book handy. Commit your favorite jokes to memory—preferably the ones you think are *mostly funny*!—and enjoy a good laugh at the expense of Mother Nature, or maybe even a meteorologist. Go ahead, we can take a joke. Take this one for instance:

A blonde, a lawyer, and an engineer walk into a bar on a rainy afternoon. The lawyer takes off an expensive, soaked trench coat and sits near the door so he can be the first on the scene if anybody slips on the steps. The lawyer watches the engineer spend five minutes pushing on the pull-open door. And the blonde, soaking wet having come in with neither a raincoat nor an umbrella, sits down at the bar and complains to the bartender she's sure she heard the meteorologist on TV say it would be nice outside today.

So the bartender says, "Oh . . . you must be new in town. That's just one of our local TV meteorologists trying to be funny. This is Seattle!"

OK, that one's only *partly funny*, and if you don't get it go spend a little time in Seattle and learn their simple rules for weather forecasting: *If you can't see Mount Rainier, you know it's raining. If you can see Mount Rainier, it's about to rain!* But I'm sure you take my point that weather can be funny. Please enjoy this book and hopefully it'll make you chuckle, groan, or maybe even laugh out loud.

Let me end with a special note of thanks to the AMS Publications Department in Boston. Its staff members' enthusiasm and professionalism were a joy and were essential in helping me develop this book. I also need to thank my friend and colleague AMS Executive Director Keith Seitter and my fellow members of the AMS Council (board of directors) for their willingness to go along with my crazy idea to do a weather joke book to raise funds for the work of the Society. We at the AMS hold ourselves to the very highest standards of professional and scientific excellence and integrity, and we believe we compromise none of that to share a laugh with our members and with you, dear reader.

—Jonathan T. Malay, AMS Past President
Fredericksburg, Virginia

Bad Weather

(If it wasn't for bad weather, we wouldn't
have no weather at all!)

A funnel thing happened to me on the way to the station today.

What happened to the cow that was lifted into the air by a tornado? It was an udder disaster.

Tornado: Nature in the roar.

I have a severe weather tip. In case there's a tornado warning, be prepared to go to a safe place . . . like Peru.

If you encounter a tornado while driving, remember the funnel has the right of way.

A really bad storm is what we need from time to time to remind us we're really not in charge of anything.

Ever wonder why tornadoes never sound like passenger trains?

One more storm and you won't have to go to the beach ... the beach will come to you.

We have a low pressure system with an attitude problem.

Today's weather will be followed by a solemn moment of profanity.

Mother Nature gives us these storms so people who aren't interested in sports will have something to talk about.

I've been afraid of this ... the environment is fighting back!

Beach erosion: The great terrain robbery.

Beach erosion: Shore Leave.

You've heard of déjà vu, right? Well, we had a vuja day ... you get the feeling you've never been through a day like this before.

Today we had a Nor'easter. I know in the spring we'll have a Nor'passover.

It was one of the worst storms we've ever had, unless your name is Dorothy and you have a dog named Toto.

I'm going to give Mother Nature a 15-yard penalty for unnecessary roughness.

If you're planning to go to the beach, don't bother, the beach is coming to you.

If you want to do something about violence on TV, ban my weather forecast.

I wonder if Chicken Little was on to something.

This is a good time to watch the ocean. The special effects are spectacular and there are no commercial interruptions.

I've often wondered about this. Do fronts have rears?

It's a good thing I ordered sunny-side-up eggs for breakfast this morning.

If you have sunny-side-up eggs for breakfast tomorrow that will be the brightest thing you'll see all day.

I can't understand why it's raining . . . the weekend is over.

There's no such thing as bad weather . . . just inappropriate clothing.

The weather was so rotten it was hardly worth calling in sick.

Light rain: Rain that's easy to carry.

Light rain: The same as regular rain but with 20% less calories.

Light rain: It's raining kittens and puppies.

I don't care what anybody says . . . this weather is better than no weather at all.

Everyone knows Sunday is a day of rust.

We have a low pressure area locked in mortal combat with a high pressure area. If you're a gambler the smart money is on the low.

Neither rain snow nor gloom of night shall stay the swift couriers from the completion of their appointed rounds. I'm not talking about mail carriers. . . . I'm talking about people who go to garage sales.

What we need today is a little Dutch boy to stick his finger in the clouds.

A lot of people enjoy walking in the rain ... except the people who have to.

I'll tell you how bad our weather's been. I met this kid who doesn't believe in three things ... the Tooth Fairy, Santa Claus, and the sun.

What's really annoying is all the people telling me "At least it wasn't snow."

Everyone says, "At least the air is free." Sure, who'd want it?

This weather is something. I never thought I'd see the day when indirect lighting was the sun.

No one ever talks about the good side of this weather. Do you realize that if it wasn't for coughing and sneezing, some people wouldn't get any exercise at all?

Tonight we're offering a special prize to the viewer who draws the best picture of the sun ... from memory.

The good news is that tomorrow's rotten weather will not be as rotten as today's rotten weather.

Keep in mind that nothing lasts forever . . . with the possible exception of this weather.

A car pool is what you get when you leave your car roof open in a rain storm.

Where is the sun? Actually the question for people with a short memory is . . . *what* is the sun?

This just in . . . tonight's rain has been cancelled on account of the game.

The trouble with today's weather is it's too true to be good.

Into each life some rain must fall. But why after you've just finished washing your car?

Into each life some rain must fall. Followed by clearing sky, a southwest wind, and temperatures in the 70s.

Into each life some rain must fall. Usually on weekends.

Into each life some rain must fall. But nobody ever said anything about a monsoon!

Into each life some rain must fall . . . especially if you leave your car windows open!

Into each life some rain must fall. But no one ever told me about the mud!

Why is it that a heavy downpour can wash away tons of topsoil but won't remove one ounce of dust from my car?

You know what happens after a dry spell? It rains.

The storm is moving slowly, weakening, and heading south . . . kind of like me.

You know those picture windows that bring the whole outdoors into your living room? Tomorrow your kids will be doing the exact same thing. . . .

Tomorrow will be mostly boring with a 50% chance of gloom by afternoon.

I'm the world's best forecaster. . . . I'm the raining champion!

It's a well-known scientific fact: Rain is caused by cold fronts, humidity, and weekends.

I would like to congratulate all the mothers in the listening area dealing with this weather. . . . You've just won a houseful of kids.

If you want a rain check for the weekend, forget it. Meteorologists don't give out rain checks.

The weather's been so bad that kids are no longer asking their parents the question: Why is the sky blue?

You've heard the expression "Let a smile be your umbrella." Well, it doesn't work. You'll wind up with a mouthful of rain.

Do you know what they call people who believe in letting a smile be their umbrella? Wet.

There's been so much rain the last few days I thought I was on vacation!

Rain—the chief product of the country you go to on vacation.

Did you wash your car again? Car washing precipitates precipitation!

This is great weather for anyone who wants to live in a car wash.

This town is the place where all the scattered showers seem to get together.

This town is a great place, but today it sure needs a roof.

The morning's rain won't last past noon. Then it'll be an afternoon rain!

This is great weather if mildew is one of your favorite plants.

Four straight days of clouds and rain and nothing to show for it . . . except mildew!

It's so humid you start to mildew immediately. In just one hour you look like a giant Chia Pet.

This is a good time to add to your mildew collection.

It'll rain for the better part of the day . . . or I should say the worst part?

If you're a person who's outstanding in your field . . . your field is going to be wet today.

We had nice weather here this week. It rained only twice . . . once for three days and once for four days.

We don't need any calendars. When it rains we know it's Sunday!

What do you call two straight days of rain? A weekend!

This weather pattern is unusual since most of the time it doesn't rain until the weekend.

This is the only place where there are four seasons and just one forecast . . . rain.

It's interesting. We have coffee without caffeine, beer without alcohol, and milk without fat. Can't they come up with a weekend without rain?

Notice the rain seems much wetter when it comes on a weekend?

Ever wonder how come the day with the least amount of sunshine is called Sunday?

Bucket seats are what you have inside your car if you leave your car windows open in this rain.

Our weather will be outstanding today. That means you'll be out standing in the rain!

There's a big storm brewing. The question is whether it's regular or decaffeinated!

The weather is rated PG . . . pretty gloomy.

I have to apologize. My forecast is going to include indecent four-letter words like snow, *rain, cold, wind,* and *brrr*!

I want to alert all cable viewers that for an extra $3 a month they can block out my weather forecasts.

The news director's daughter was getting married, so the station's weather department decided to give her a shower.

This weather has its good side. It gives people without kids something to complain about.

How come the threat of showers postpones more yard work than golf?

This weather is really messed up. I had to wash my car three times to make it rain!

The weather should improve by 1:30. Good thing January 30 isn't that far away!

I saw an ad for a special sale on umbrellas . . . but only at precipitating dealers.

This is bobsled weather. It's going downhill fast!

The sun didn't come out today. Well, would YOU come out in this weather?

Today you're going to see what happens when you don't tip your meteorologist!

If the rain bothers you, take two umbrellas and call me in the morning.

Let's go to the 40 days and 40 nights forecast.

My forecast says we're due for more of the same: wind, fog, rain, and scattered complaints.

How come when sheep get wet they don't shrink?

You can look at the good side of this weather . . . it keeps the dust down!

One good thing about this weather . . . you're saving on sunscreen!

The rain came just in time. For a while I thought we were going to run out of mud!

We need this forecast like a turtle needs a speedometer . . . like a mermaid needs pantyhose . . . like a fish needs running shoes!

You're in luck, folks: I was told in order to improve our ratings I'd have to forecast some better weather.

So you think you have a rotten job? I have to stand here and tell you it's going to rain again this weekend!

Don't complain about the weather. It'll find a way of getting even!

The weather is so bad people are renting movies with a lot of sunshine, like *Lawrence of Arabia*!

I miss the sun. I'm getting suntimental!

I was in the weather center monitoring this weather. Of course that is the only place to monitor this weather. I'm sure as heck not going outside!

Where's Gene Kelly when you need him?

Global warming has become global storming!

We broke the record today. It was the record for the number of people who complained to me about the weather.

Everyone talks about the weather. And this week it deserves everything they say about it!

The wind is blowing from east to west and the rain is falling from up to down.

If it looks like rain ... it's probably water.

Rain and Monday. It's the old one-two punch!

Sorry all you golfers ... it's time to give the grass on the course a rest.

Why am I standing on these newspapers after walking on the flooded streets? Because these are the *Times* that dry men's soles.

After massaging the patient's back, the chiropractor said, "I think it's going to rain." "What makes you say that?" asked the patient. "Well I can feel it in your bones."

In some ancient civilizations they worshiped the sun. We might too if we ever saw it.

Knock knock.
Who's there?
Dwayne.
Dwayne who?
Dwayne is leaking through the roof! It's been a long week ... one damp thing after another!

I was dining in a very upscale restaurant last night. I asked a waiter, is it raining outside? The waiter said, "Sorry this isn't my table."

It's really annoying when you're wearing a raincoat and carrying an umbrella and the clerk at the store says, "Have a nice day."

Around here, there's a technical term for a sunny, beautiful day that follows a rainy period. It's called Monday.

On a rainy Sunday, churchgoers were struggling at the entrance with their umbrellas. A family of four arrived with a huge umbrella and the usher asked, "Is that a golf umbrella?" "Yes," the father replied, "may we pray through?"

We just ordered a new designer Doppler radar. Rain areas are mauve, snow is taupe, and tornadoes are puce.

Hips, joints, knees, lower back ... I guess my personal Doppler says, "It looks like rain."

The rotten weather will continue. Another day, another Doppler.

All you need is a Doppler and a dream!

Morning rain: A rush shower!

Coincide: What you do when it rains.

Roofer: A person who's shinglin' in the rain.

It's a rotten morning. For breakfast you could be having drench toast.

Ringing wet: What you get if you leave your alarm clock out in the rain.

Umbrellas: Things that are shipped from France by parasol post.

You've heard the poem "Rain rain go away. . . ." This must be that other day!

Football is a sport that's played in almost any kind of weather. A few years ago, the home team captain was faced with a Sunday of torrential rain. The field was a quagmire as the official tossed the game-opening coin. The home team lost the toss, and the other team's guy said they wanted the ball. "Do we really have to play in this flood?" asked the home team captain. "Of course," said the game official impatiently. "Now which side do you want?" So he replied, "In that case we'll kick with the tide."

Kangaroo mother to another: "Don't you just hate these rainy days when the kids can't play outside?"

Hydroelectric dam: What you'd hear if a person uses an electric shaver in the rain.

Mean rainfall: If it falls even on the Easter Parade!

Rain dance: An element of chants.

Permutation: A very bad hair day.

Rain: Something that when you take your umbrella, it doesn't.

Our government created these three-day weekends because they couldn't cram all the rotten weather into just two days.

Clouds

(I can't see the weather . . . the darn
clouds keep getting in the way!)

We always seem to be in the partly that's cloudy.

Partly Cloudy: A few clouds, but less than mostly and
more than moderately . . . which all things taken together is about the same as partly sunny.

Every cloud has a silver lining . . . but those are the better clouds. The ordinary clouds come in polyester!

Remember every cloud has a silver lining . . . but then
the price of silver is down these days.

The difference between partly cloudy and mostly sunny
is you never see mostly sunny at night.

We have some broken clouds . . . but we'll try to fix them before the news is over!

To bring our ratings up we're going to show risqué cloud formations!

Have you ever heard of the terrible effects of cloud burn?

The clouds are up. The question is: What are they up to?

Many clouds are shaped like animals, but the sheep is the only animal shaped like a cloud!

The three basic cloud types seem to be bunny rabbit, cotton candy, and a snowman.

Partly sunny followed by partly cloudy is bound to be partly accurate!

Two clouds were having tea and one said to the other, "Shall I pour?"

It's my fault . . . this morning I ordered my eggs cloudy side up!

Changeable sky: Stratus various.

Overcast: Cloudy duty time.

Cloud observers: People who take their job cirrusly.

Fog

(Let's make this perfectly clear . . .)

There's a song written about the fog: "I wonder who's kissing me now?"

Did you ever try to build a fog man?

Did you hear about this guy who tried to kiss his date in the fog . . . and mist?

There will be no weather tonight because of the fog. I couldn't find the forecast!

You know those traffic flow signs on the coastal road? I saw one that said, "If you can't see this sign, please pull off the road!"

What's green and makes a loud noise? A frog horn!

When you go outside you won't have the foggiest idea where you are.

Talking about fog: I was halfway to work this morning and I realized I wasn't even in my car!

If you're going to the game tonight, all seats will have an obstructed view.

Fog: The air apparent!

Extremist: A dense fog!

A motorist driving in a dense fog was confidently following the taillights of a car ahead for a full hour. Suddenly the lights on the lead car stopped and the two cars collided abruptly. "Hey, why don't you signal when you're going to stop?" yelled the driver of the second car. The first driver answered, "Why should I? I'm in my own garage."

Do you have fear of overcast skies? Then you're suffering from cloudstrophobia!

What happens when the fog dissipates from Southern California? UCLA. (Ron Gird)

Dry

(A little dry humor, anyone?)

It's been so dry . . . I had to wash my car three times to make it rain!

To draw attention to the drought, tonight's TV movie was changed from "How Green Was My Valley" to "Lawrence of Arabia."

Washing your car to make it rain doesn't work.

To alleviate the lack of rain people are taking fewer showers. Showing the movie *Psycho* is really paying off!

A drought is when it's bottoms up for ponds, streams, and reservoirs.

I want to announce the weather department is planning a rain dance . . . weather permitting.

Because of the drought, Jack and Jill were only able to fetch a pail of water on alternate Wednesdays.

Rain, rain, come and stay!

This is a good time to paint your house brown, so it matches your lawn!

A man was given a summons for watering on a wrong alternate day: He was caught wet handed!

I have a plan to alleviate the drought: If farmers can get paid not to grow crops, why can't I get paid not to take showers?

It's been so dry lately, I don't water the plants . . . I dust them!

It's true there's still a drought in the Southwest. I just received a letter from a friend who lives out there and the envelope was sealed with thumb tacks.

Jack and Jill went up a hill to fetch a pail of water. They both received a summons because they took more than they oughta!

The lack of water is causing problems. A porch caught fire and the fire department had to blow it out!

Where is the rain? Actually, the question for those with a short memory is "What *is* rain?"

I finally found a diet that works. . . . You eat only when it rains.

It should rain by 8:30. August 30 isn't that far away!

I just thought of a great idea for conserving water: You dilute it.

The drought is so bad the management had to drain and close two lanes of the swimming pool.

It hasn't rained for such a long time I've forgotten where the little knob is that turns on the windshield wipers in my car!

I like the dry weather. I use the dust as a protective coating for my furniture.

Dry? I was in a seafood restaurant last night and the lobsters were sitting on a couch waiting to be picked!

The drought is so bad my plastic plants died!

It is so dry I went to a restaurant and they gave me a glass of H_2. . . . There was no O in it!

It's been so dry people are having mirage sales!

To conserve water, we recommend you brush every other tooth, or maybe brush your upper teeth on even-number days and lower teeth on odd-number days.

It's been so dry my grass comes in regular and extra crispy.

It's been dry. I was at this restaurant and the bartender came over and asked if I was done with my ice cube!

Because of the drought, you can water only on Tuesdays and Thursdays. The fire department can only put out fires on Wednesdays, Fridays and Saturdays!

A successful rain dance depends on an element of chants!

Arizona: Where a baby's first words are, "But it's a dry heat."

Drought: When you go from one ex-stream to another.

Desert . . . long time no sea.

Fall

(Colorful jokes!)

I'm puzzled. What happened to the leaves before Isaac Newton discovered gravity?

It's the first day of fall. I guess I can give up having a good summer.

It's the first day of autumn and two things begin today: fall and complaining about the cold.

I love fall in the city. . . . You can watch the colors change from green to gold to brown. And that's just the air!

Trees are really amazing plants. . . . The way they can lose many more leaves in the fall than they grow in the spring.

The question is: What will fall on your lawn first, an autumn leaf or a holiday gift catalog?

I always enjoy autumn leaves, especially when autumn leaves them on someone else's lawn!

It would be a lot easier if the birds would stay here for the winter and the leaves on your lawn would go south!

Modern life in autumn: You pay a kid to rake your leaves so you can get to the health club to work out.

This is a nice time to take a long ride into the city and watch the colors turn gray.

It's the time when the colors outdoors turn red, orange, and yellow. But I have things in the refrigerator that do that every week!

I use a leaf blower on my lawn. It's called the wind!

The days are getting shorter, which is OK since I always thought 24 hours were too many.

The days are getting shorter and the clothes longer.

October is a happy time of year. You get up in the morning and there's nothing to mow, rake, or shovel.

October is a beautiful month to take a hike . . . or tell someone to take a hike. It's also a beautiful month to fly a kite . . . or tell someone to go fly a kite.

October is the happy side of summer and the pleasant side of winter.

October is when greens turn to browns and tans turn to pales.

Enjoy these few days between lawn mowing and snow shoveling.

I'm planning ahead. I'm starting to write my new "How cold was it?" jokes.

I'm starting to hate winter now. I want to avoid the rush!

You can tell winter is approaching . . . the kids are wearing much heavier clothes to surf the Internet.

It's November . . . leaves are falling, days are getting shorter, and a bird in the hand is the Thanksgiving turkey.

November is when leaves start falling and fall starts leaving.

November is when leaves go from flash to trash.

Hey kids, sorry but they don't close the schools for falling leaves.

November is when you go from leaf watching, to raking, and then to aching.

The winds were gusty . . . due to the simultaneous use of leaf blowers by everybody in town.

Fall is when the weather goes from lovely to shovely.

It must be the football season . . . my end zone is frozen!

I saw the first sign of winter today: a line of senior citizens at the post office picking up change of address forms.

Indian summer is a beautiful spell of weather brought on by putting away all your lightweight clothes!

You can see it's just past Indian summer when you see a convertible with the top down and the driver is wearing ear muffs and mittens.

This is more like "election day summer." We're getting a lot more hot air than expected!

Humpty Dumpty sat on a wall. Humpty Dumpty had a great autumn!

October is a good time to take your electric blanket in for a tune-up.

It's a sad time of year when the trees shed their leaves and the baseball teams shed their managers.

It was so cold the local pro football team went into a huddle and didn't want to come out.

It's a typical autumn day. The frost is on the pumpkin and the linebackers are on the quarterbacks.

That dance the football players do in the end zone is really fascinating. I can't tell whether they're excited or trying to bring on rain!

Cold? I saw some autumn leaves break when they hit the ground.

Today is the first day of fall. That means we can stop sweating from the heat and start sweating from the heating bills!

You're familiar with the greenhouse effect? During football season it's called the doghouse effect.

It's fall when summer's suntan turns to winter's frostbite.

It's that time of year I'd feel much better if it wasn't that time of year!

Temperatures are falling. When they reach 40° . . . sell!

Do you know you can get the long-range weather forecast by watching the World Series? The announcer always says, "For the team that loses it's going to be a long, tough winter."

This is the time of year weather prophets look at the woolybear caterpillar to see if we're going to have a severe winter. That's known as the "furcast."

I think we're going to have a tough winter. I just saw a squirrel burying canned goods.

Don't you think this is a good time to weather strip our border with Canada?

Today you go outside to watch the leaves drop. I'm going to stay in the weather center and watch the temperature drop.

I'm saving money on heat this year. I've converted my furnace to burning gift catalogs.

One of the nicest things about December is that January isn't here yet!

I just got my car ready for winter. I caulked the windows, installed a bigger muffler, and put hot chocolate in the radiator.

It's November (fill in the date X). That means only (25 plus 30−X) days left of people asking weather forecasters if we'll have a white Christmas.

Winter must be getting close. At the airport I saw a flock of golfers heading south.

Fall is a season for big decisions . . . like whether it's too late to start spring cleaning.

The landscape gets prettier in the fall. Most of the local residents go back to wearing long pants.

It's frightening. Winter's almost here and we still have potholes left over from last winter!

I know winter is getting close. I just saw my first Zamboni!

Question: Can we start using some of that daylight we were saving all summer?

Fall is the time of year Mother Nature stores up all the nuts you'll see on the highway next spring.

First frost: When you're caught with your plants down.

Fall is the season women buy winter clothes so they can have something to wear when they go shopping for their new spring outfits.

The birds are starting to migrate south. That's poultry in motion.

The best way to see fall foliage is in your autumn mobile.

Weather stripping: An exotic dance performed outdoors.

Fanatic: Where you store your fans for the winter.

Fall foliage: An Oct. of God!

This is Indian bummer!

Remember the old weather adage: Rain in November, Christmas in December.

It's a miserable day, but it's hard to feel blue when there's so much red, orange, and yellow around!

Fall

It's the time of year the days are getting shorter. Except for the ones where you have to wait for the appliance repairman.

Into each leaf-clogged gutter some rain must fall.

That's an expensive store! People are shopping hurricane style: They're spending 70 to 80 dollars an hour with gusts to 105!

Hurricanes

(If you can laugh in the face of a hurricane,
you're probably in very big trouble!)

In a hurricane there's nothing to fear but atmosphere itself!

Notice how anytime you have a mix of male and female hurricanes there's a lot of little baby hurricanes popping up?

When the wind exceeds 74 mph they're called hurricanes ... and other names!

It's the hurricane season ... the only time you can get a ticket for going more than 75 mph and never leave your house.

It's the hurricane season ... when your garage door stays down and your house goes up.

Hurricane tracking is Mother Nature's way of teaching us geography.

They give hurricanes cute, wimpy names. What they need are names that are synonymous with disaster and destruction . . . something like Hurricane Internal Revenue Service!

How about Attila the Hurricane?

A hurricane is a tropical storm with an attitude.

A hurricane is interesting because the storm has an eye at its center and the word also has an "i" at its center.

Before a hurricane, you find supermarkets full of people who don't want to be caught with their pantries down.

With this hurricane, I'm going to give Mother Nature a 15-yard penalty for unnecessary roughness!

The first question people ask after a hurricane is: "If a tree falls in your backyard and no one is there to hear it, who is responsible for getting rid of it?"

Before the last hurricane in Florida, the roads were jammed with people trying to escape . . . and those were just the insurance agents!

Everyone knows the best place to be when there's a hurricane is in a state that's not having one!

They measure a hurricanes force by how many TV reporters are blown out to sea.

Some of the most dangerous things about hurricanes are flying debris, flying trees, and flying TV reporters!

Today's beauty tip: In a hurricane the average hair spray doesn't work!

The hurricane blew my car away and left another one in its place. It must have been a trade wind.

After the last hurricane hit, a man called his insurance agent and reported, "My house is a split level. . . . Of course it didn't start out that way."

This just in: A tropical storm is headed for Florida. The National Weather Service said they wouldn't declare it a hurricane until all the insurance companies cancel their policies.

An insurance agent suggested to a client who had fire insurance on his business that he should also be covered for hurricanes. The client asked, "How do you start a hurricane?"

This tropical storm may hit Florida. You can tell by the flock of plywood salesman heading south!

The bad news is that a hurricane with 140-mph winds is approaching Florida. The good news is that cars going north are getting 400 miles per gallon.

Eye shadow: A person who secretly keeps watch over hurricane movements.

Hurricane: What Able said to his brother when he was late for school.

Walt: How have you been since the last big hurricane? Jim: I've been living in a mobile home. Walt: Oh, you moved? Jim: No, before the storm it wasn't mobile.

An old rule of thumb in TV journalism says: If you see a reporter standing outside in a hurricane, he or she is not a senior member of the news team!

A meteorologist named Norm
Predicted a tropical storm
When the sun kept shining
He started his whining
Which wasn't in very good form.

Extratropical low: A spare low kept handy, just in case.

Political Season

(These jokes practically
write themselves!)

This is the time of year the big problem is the wind chill factor. The candidates supply the wind and we get the chills!

Has anyone ever looked at political speeches as the possible cause of global warming?

Scientists are worried about the greenhouse effect and the impact it will have on our lives. I'm more worried about the White House effect!

I never vote for anyone dumb enough to be out campaigning in this weather.

He could win by a mudslide!

All the exit polls predicted I'm going to have a good forecast.

I heard many of our past presidents were meteorologists. Isn't that why they're always greeted with "hail to the chief"?

I call this "election day summer." We're getting a little more hot air than usual.

The important thing to consider in this election is if you are better off with today's weather than the weather you had four years ago.

The weather does as it pleases. Wouldn't it be great if it was something we could vote on?

Political speeches are a lot like a 20-inch snowfall. It's deep, affects everybody, and six months later you have no clue it even happened!

My forecast has just received bipartisan endorsement.

It was so cold in Washington this week that I saw Republicans and Democrats hugging each other!

The icy chill was coming off a cold front between the House and the Senate.

Notice how the strong winds die down right after an election?

I'll take credit for this weather ... before some politician does!

Halloween

(Trick or joke!)

It was raining so hard the kids had to go trick or creek.

I don't mind if I have Halloween candy left over. It comes in handy for caulking the windows!

In honor of Halloween, my seven-day forecast comes from ghost writers!

There's not even a ghost of a chance of any rain on Halloween.

My forecast gives the precise times of sunrise and sunset. It's a public service for vampires.

I don't know which is scarier . . . Halloween or my forecast!

Thanksgiving

(Careful . . . these jokes are filling!)

I have to apologize. . . . Today's forecast is made up entirely of leftovers.

This morning the Thanksgiving turkey came out and saw its shadow. That means we'll have six more weeks of leftovers.

Thanksgiving is so cold you'll see a turkey with a capon!

This Thanksgiving you may want turkey, but you'll get chilly instead!

Turkey is not the only thing left over from Thanksgiving. How about this weather?

For Thanksgiving the rain will wet your appetite!

I'm calling for a weight watch and an indigestion alert.

The only thing keeping people warm is their heartburn.

The weather diet: You can lose up to 15 pounds with gusts to 25.

It will be raining cranberries and drumsticks.

Indian to Pilgrim: "Are you serious . . . you want to have an outdoor picnic in late November?"

Today is Black Friday, a great day for shopping with a high of 47 and a low of $39.99!

For Thanksgiving, we're having an eat wave.

Floods

(Water you wading for?)

The only thing worse than a flooded basement is a flooded attic!

The highway authority called to remind everyone that traffic rules apply, but homes have the right of way.

Because of the tremendous rain, crime took a new twist in the area . . . piracy!

You know it's too wet to play golf when your golf cart capsizes.

This season has been so wet I saw this bumper sticker on the highway: "I brake for fish."

What a downpour! If your kid just had lunch, make him wait an hour before going out to play.

Residents are warned not to leave their house for at least an hour after eating.

It's been raining so hard my wake-up service said "rise and float"!

What a storm! Last night at the mall I saw people contributing to the Salvation Navy.

We'd better start to gather the animals two by two. . . . But this time we're not taking two mosquitoes!

Into each life some rain must fall. But at least Noah had enough sense to build an ark!

The skies opened up and the rain poured down. Noah looked up and said, "I knew it was a mistake to wash the ark this morning!"

The marine forecast calls for seas five to seven feet on the ocean, three to five feet in the harbor, and two feet in your basement.

The sudden rise of waters along the shore introduced landlubbers to the pleasures of boating.

Mother Nature says, "This flood's for you."

I asked this guy, "How do I get to the middle of town?" He said, "It's about five miles as the crow floats."

Because of the flooding on the highway, traffic was tied up for hours. The good news . . . the fish were biting!

Announcement to those stuck on the highway in high water: "Your seat cover can be used as a flotation device."

This is what happens when too many drips get together!

I saw this bumper sticker on the highway that said "I'd rather be sailing." And a few minutes later he was!

Did you hear about the flooding on the expressway? It's water under the bridge.

What goes drip-drip instead of tick-tock? A flood watch.

It's been so wet here that a man's home is not only his castle but his boat too!

Lots of the area was flooded. What a water-revolting development!

Flood: When a river gets too big for its bridges.

This sign was seen at a Federal Loan Office in a flood disaster area: LOANS MADE WHILE YOU WADE.

Flood control: Measures taken to control floods that are much more effective in dry weather.

Half the valley was flooded by the worst rainstorm of the century. Desperately, a man called the rescue hotline: "Help me! I'm standing in two feet of water." The phone dispatcher said, "There are people around here that are worse off than you are." The man said, "But I'm calling from the fourth floor!"

During a recent flood a farmer whose home was washed away discovered he had musical talent. "My wife held on to a table and floated down the canyon," he said to a friend. "But how does that prove you have musical talent?" the friend asked. The farmer said, "I accompanied her on the piano!"

Forecasting

(When you get paid for guessing!)

Weather forecasting is a kind of work no one notices until you do it wrong.

My forecasts are never wrong. . . . It's the weather that's wrong!

At least I'm right about one thing. . . . There's some kind of weather every day.

I think the reason my forecasts have been a bust is I've been looking too much at the camera instead of out the window.

We had puddles of partly cloudy.

You have to be careful with any forecast that starts out with once upon a time"!

My forecasts have earned the Seal of Approval of the American Apology Society.

It's at the National Weather Service where many a true word is spoken in guess.

Weather forecasting is not an exact science. When Noah got on the ark, the meteorologist that day predicted partly cloudy.

I wasn't able to clear my driveway today because I needed my shovel to prepare my forecast.

I actually predicted partly sunny. . . . The sunny part was 200 miles away!

I climbed the ladder of success wrong by wrong.

We must be having a problem with air pollution. After my last forecast, the news anchors were all holding their noses.

There's a meteorological word that describes my forecast: *Oops*.

I spent a long sweaty day shoveling off a light dusting.

The way things are going, they're going to add laugh tracks to my forecasts.

I usually predict the right weather but seldom on the right days.

I wouldn't be wrong so often if I didn't limit my forecasts to the future!

Computer Forecast: The dot.com before the storm.

You thought the basketball playoffs lasted a long time. . . . Wait until you hear this forecast!

You can rest assured when you wake up tomorrow morning there will be weather.

This is one of the most difficult months to forecast. The others are January, February, March. . . .

I want to alert all cable viewers for only an extra $8 a month you can descramble my extended forecast.

I just got my forecasts ready for high-definition TV. From now on, my forecasts will be much clearer.

I try to forecast one day at a time. Right now I'm working on last Tuesday!

People have been asking me to try to get the weather to match my forecasts.

Today is Friday the 13th. So if my forecast is wrong, I have something to blame it on.

Just remember my forecast is an estimate. . . . Your weather may vary.

I'm going to figure out tomorrow why the things I predicted yesterday didn't happen today.

And now for my three-day forecast: Saturday will be followed by Sunday, and then by Monday.

My forecast was for a morning low of 26°. A woman called and asked me if it will be freezing. I replied, "Anything below 32° is freezing." She said, "I didn't know. I'm new in this town."

Doctors and forecasters follow the same belief: It's a science if they're right and an art when they're not!

We're very cautious with our forecasts at this news channel. Yesterday I said there was 70% chance of Tuesday.

My 90-day outlook calls for an equal chance of the temperature and precipitation values being above or below normal.

Extended forecast: Fooorrreecaaasstt.

Weather forecaster: A person to whom one and one are two . . . probably.

Weather forecaster: The only profession whose work depends on the weather, but who gets paid rain or shine.

One of the most important things about being a TV weather forecaster is honesty. If you can fake that, you've got it made!

There's a 100% chance of rain or no rain tomorrow.

A meteorologist was given a ticket for making an illegal U-turn and had to appear in court. The judge, after hearing all the facts, told him: "Fine today and cooler tomorrow."

A new student at a state university spent a day on campus orienting himself to the layout of the buildings and classrooms on his schedule. He was elated after reading a sign posted in one hallway: "There's a 75% chance that you have reached the Meteorology Department."

Forecast for this weekend's golf tournament: Golf balls the size of hail stones.

The local forecaster's predictions were wrong so often that he was embarrassed and applied for a similar job in a different part of the country. The job application called for the reason he left his previous position. He wrote in the blank line, "The climate there didn't agree with me."

A tourist stopped at a general store in the backcountry. He noticed an old-timer sitting in the sun holding a piece of rope. The tourist approached the old-timer and asked, "What do you have there?" "This is a weather gauge, young man," the old man replied. "How can you possibly tell the weather with a piece of rope?" the tourist asked. The gent smiled and said, "It's simple sonny. When it swings back and forth it's windy and when it gets wet, it's raining."

Spare me your map. Your high your low;
Don't tell me how the winds will blow;
Just work it out and let me know;
Will it snow? Yes or No!

A local meteorologist was amazed to find that a blood vessel on his leg enlarged every time the air pressure dropped. He started predicting using his weather vein.

A university mathematics professor was intrigued by the National Weather Service's forecast that mentioned "the chance of rain as 7 in 10." He phoned and asked, "What theory of probability was used to arrive at your numbers?" The weatherman said, "Sir, we've got 10 guys working in this office. If 7 say it's going to rain and the other 3 say they're crazy, we've got our percentage."

The easiest person to forgive is the meteorologist who forecasts a blizzard and is wrong.

The TV News Weather Center was being renovated, and this sign was posted on the wall: WET PAINT. BE-COMING DRY LATER IN THE DAY."

Seersucker: A person who believes in forecasts made by weather prophets.

Summer Forecast: "Thunderstorms, followed by blinking digital clocks."

Good Weather

(Probably the least funny chapter in
this book.)

Do you think if Ben Franklin waited for a day like today
to fly his kite he would have discovered solar energy
instead?

Tomorrow will be sunny and nice . . . weather
permitting!

I just saw my mail carrier stretched out relaxing out on
a lawn. He said, "I can handle rain, sleet, and snow but
with this great weather . . . forget it!"

Postal workers are not delayed by rain, sleet, snow, or
gloom of night, but on a day like this they head for the
golf course!

The day will be bright and sunny . . . and the night a lot darker!

The air quality is going to be excellent today so go ahead and breathe every chance you get!

Fresh air . . . it's one of the few substances not yet available in artificial form.

We finally got some fresh air. So you can now resume breathing.

Believe it or not, there are people who look out the window on a day like today and say, "It's too beautiful to stay inside . . . let's go to the mall!"

This weather is so beautiful . . . at the zoo I saw a turtle with the top down!

Today it was 79°, but if you factor in the winds it felt like 78°!

We should be thankful the IRS isn't in charge of our weather. Today is going to be so beautiful they'd find a way to tax it.

I love this weather . . . blue sky, mild temperatures, and the wind blowing through my ears!

In this weather I like to stick my head out of my car window, look up, and smile for a satellite picture.

With this weather I've come up with a new concept in meteorology . . . tipping your meteorologist!

I saw this bumper sticker: "Have you hugged your meteorologist today?"

Today will be a great day for people to reset their sundials!

Now that I've finished straightening out the weather, I'm going to work on the economy, crime, and traffic congestion!

Personally I prefer bad weather. I hate it when the barometer is steadier than me!

It's a great day to fly a kite, but some kids will wonder why you have to run with it and why it doesn't come with batteries!

We finally got some good weather. People are opening their car windows and roofs to let the mud out.

Today's weather is Mother Nature's way of telling us to go out and play.

Today's great weather is Mother Nature's way of apologizing for February.

We have 3H weather: Hip hip hooray!

We have 3H weather: It's good for hot dogs, hamburgers, and hammocks!

With this great weather, it's a great day for the race . . . the human race!

It's the kind of day you need a sweater to keep warm and sunglasses to look cool.

The best antidote for rotten weather is a Monday!

The weather is so dull the tide's going out and not coming back in. (Attributed to Fred Allen)

Tomorrow will be rain free . . . about the only thing that doesn't cost anything around here!

I'm getting a suntan just thinking about this forecast!

Today will be a great day. . . . We'll have six inches or more of hail-sized golf balls! (Jon Malay)

Great weather yesterday: It was shining cats and dogs!

A big high pressure is coming out of Canada. So why can't we have our own high pressure systems? Why do we have to import them?

You get up every morning, the sun is shining, the birds are singing . . . and there's just too much pressure for a barbecue!

I just thought of a new business: Replace a beautiful, bronze tan with a pale look in just a few minutes. It's for people who called in sick and then went to the beach.

It's amazing . . . I saw golf balls the size of hailstones!

It was 74° and the air was saturated with golf balls.

It was a perfect spring day. The sun was shining, a nice breeze was blowing, birds were singing . . . and the lawn mower was broken!

Why does everything happen on cloud 9? What happens on clouds 1 through 8?

If you're out in the sun tomorrow be careful not to get too close to it!

Anyone who complains about this weather should live in a mall.

This just in . . . I'm having a good hair day!

It will be bright and clear. That's the weather . . . not me!

Don't you hate it when the weather is nice and there's nothing to complain about?

It was so nice today, on Wall Street they were doing outsider trading!

We're calling for just a few broken clouds. But we'll try to fix them before the news is over!

It's a beautiful evening . . . crisp, a bright moon, and everywhere you go you can hear the delicate chirping of cell phones.

If you didn't like today's weather, you're inherently cranky.

The weather is so nice my kids are fighting to see who will take out the garbage.

Did you know the sun is powered entirely by solar energy?

The weather has been very tranquil. It's a case of global yawning.

I'm predicting clear blue sky. It's azure thing!

What we're experiencing with all this sunshine is global tanning.

Question: When I drive on these sunny days I can't see because the sun gets in my eyes. What should I do? Answer: Drive on cloudy days!

If Noah had a day like today, instead of an ark he would have built a BBQ grill!

Groaners

(Hopefully, the funniest chapter
in this book!)

Why should we hope that the rain keeps up? So it won't
come down!

Where do Alaskans put their money? In a snow bank!

What kind of accident policies do skiers get? Snow fault
insurance!

What's the forecast for Mexico? Chili today, hot tamale!

What do you step in after it rains cats and dogs?
Poodles!

What actor is like a thunderstorm? John TraBOLTa!

What usually happens after a big flood? The river gets too big for its bridges!

What comes after a snowstorm? Shovels!

What does a "tempest in a teapot" mean? A storm is brewing!

What's the difference between the land and the ocean? The land is dirty and the ocean is tidey!

How does the wind blow in the spring? Easter-ly!

What's worse than "raining cats and dogs"? Hailing taxi cabs!

What do you get when you combine cold cuts with a tornado? A tongue twister!

Why did the lady go outside with her purse open? To see if there would be any change in the weather!

How did you find the weather on your vacation? I just went outside and it was there!

What did one little rain drop say to the other little rain-drop? My plop is bigger than your plop!

What's the color of waterdrops? Watercolors!

If April showers bring May flowers, what do May flowers bring? Mayflowers bring Pilgrims!

How were three fat men able to stand under one umbrella and not get wet? It wasn't raining!

What's the difference between an iceberg and a clothes brush? One crushes boats and the other brushes coats!

What's the difference between a winter day and a boxer who is down for the count? One is cold out and the other is out cold!

What's the difference between a rain gauge and a bad fielder? One catches drops and the other drops catches!

Fall: When leaves turn and terns leave.

Who was the first to come out of the ark when it stopped raining? Noah? Nope . . . the Bible says, "Noah came forth."

Teacher: How do you spell rain? Student: R-a-y-n-e. Teacher: That's the worst spell of rain I've seen in a long time.

The beach is where you get tanned by the sun and sand by the ton!

What kind of umbrella does a Frenchman carry when it's raining? A wet one.

Do you think it will rain? It depends on the weather!

Is it raining outside? Does it ever rain inside?

It's raining, let's hurry. If we hurry, will it rain less?

Did you go swimming in the hot sun yesterday? No, I went swimming in the water.

What do you get when you leave a puppy out in the cold rain? A frozen pup-sicle.

What does a snowman eat for lunch? Cold cuts.

What did one cloud say to the other cloud? Today is Sun day!

Why did I have more snow in my yard than my neighbor? Because I have a bigger yard!

Which travels faster, heat or cold? Heat, because you can catch cold!

Do you know how to make antifreeze? Sure, hide her pajamas.

Did the hurricane damage your house? I don't know. We haven't found it yet!

Why was your letter damp? Postage dew, I guess!

Where's the other windmill? We only had wind enough for one, so we took the other one down.

What did Ben Franklin say when he discovered electricity in lightning? Nothing, he was too shocked!

Do you like cycling with a group? "No, I prefer to cyclone."

Why is a crash of thunder like a jeweler? They both make the ear ring.

Why do tall people make better weather forecasters? Because short people are the last people to find out if it's raining!

When is a boat like a mound of snow? When it's adrift.

What happened when the man bought snow tires? They melted before he got home!

What is the best way to kill time in the winter? Sleigh it!

What did one neighbor say to the other after he finished building his igloo? "That's an ice house you have."

What happened when the Alaskan girl had an argument with her boyfriend? She gave him the cold shoulder.

What do they sing at an Alaskan's birthday party? "Freeze a jolly good fellow."

What did the ground say to the rain? "If you keep that up my name will be mud."

Which is heavier, a half moon or a full moon? A half-moon because a full moon is lighter.

Why should you never ski on an empty stomach? Because snow works much better!

What can pass between you and the sun without making a shadow? The wind.

What is most useful when it is used up? An umbrella.

Why is a dog dressed more warmly in the summer than in the winter? Because in the winter he wears a fur coat, and in the summer he wears a fur coat and pants.

Why is an icicle like a duck? Because they both grow down.

What happened when a couple tried to kiss in the fog? They mist.

Why don't Alaskans pay their taxes? Because their assets are frozen.

What happened to the Alaskan who was stabbed with an icicle? He died of cold cuts.

Why do mushrooms always come out after it rains? When it rains it spores.

Who inherits the wind? The air apparent.

Why would anyone shimmy up a palm tree in the noon day sun? For a hot date.

What's the difference between partly cloudy and mostly sunny? It's never mostly sunny at night!

Dad, why did the meteorologist look at the sky on Christmas Eve? To see the rain, dear.

What does a sprinter like to do during the Christmas season? He likes dashing through the snow.

What time of day do ducks get up? At the quack of dawn.

What did the Alaskan say to a girl he met at the saloon? What's an ice girl like you doing in a place like this?

Why are you wearing half your sunglasses? Because the forecaster said it would be partly sunny.

What did the snow say to the wind? Did you catch my drift?

Why kind of lights did Noah have on his Ark? Floodlights!

Why are birds unhappy in the morning? Because their bills are over dew.

What do you see in the sky on April 1? A fool moon!

What do you get when you cross a Muppet with the mist? Kermit the fog.

What did the alarm clock say to the rain? Stop, I'm ringing wet!

What does a witch cast on Halloween night? A cold spell.

Do you know what they do in Buffalo when it snows? They let it.

What's the difference between an oceanographer and an astronomer? An oceanographer cares more about the ocean bottom than the moon's behind.

What's the difference between air and water? You can make air wetter, but you can't make water wetter.

Why does lightning shock people? Because it doesn't know how to conduct itself.

Why did Ben Franklin discover electricity? Because his wife told him to go fly a kite.

What do two oceans say when they meet? Long time no sea.

How can you determine the height of a building using a barometer? Lower the barometer from the roof on a string and then measure the length of the string.

Does the ocean have dates? Yes. It goes out with the tide.

Is there precipitation on the moon? Only when it's waning.

Why did the moon go into a cloud bank? It was down to its last quarter.

Do you hunt bear? Not in cold weather!

What is the source of the Rhine and Rhone Rivers? The Rhane!

Why did the surfer cross the bay? To get to the other tide.

What do you call a frozen policeman? A copsicle.

Astronomer (looking into his telescope): "It's going to rain." Student: "Why are you saying that, professor?" Astronomer: "Because my arthritic knees ache."

How do they start the baseball season in Chicago? The mayor thaws out the first ball.

Why is the moon more important than the sun? If we didn't have the moon it would be too dark to see at night, but the sun shines only during the day when we don't need it.

What do you call a parrot in a raincoat? Polyunsaturated!

Did you hear about the performing dolphins? They got wave reviews.

How does the sun affect weight? It makes daylight.

Did you hear about the flood? Oh well, it's water under the bridge.

What did one hurricane say to the other hurricane? I've got my eye on you.

Why did the clouds separate? Because every time they got together they'd have a stormy relationship.

Did the TV meteorologist say it was going to be foggy? He wasn't too clear on it.

Why did the forecaster take so long to get dressed? He couldn't find his wind socks.

Allergy season? It's nothing to sneeze at.

Allergy season? It's a sinus of the times.

What do bees say on hot days? "Swarm, isn't it?"

How do you feel about the smog? I usually get choked up about it.

How long do you think this rain will be coming down? Until it hits the ground!

How do Alaskans dress? As quickly as possible!

Why should the air in New York City be clean? Because it has a lot of skyscrapers.

When do barbers do most of their work? Daylight shaving time.

Where did animals play video games during the great flood? Noah's arc-ade!

What do you call a heat wave in Montreal? Canadian bakin'!

Why is it so wet in England? Kings and queens have always reigned there.

What did the iceberg say to the Titanic? "Sank you very much."

Did you hear about the world's best forecaster? He's the raining champion.

How do we know that a cat likes bad weather? Because when it rains it purrs.

What goes drip, drip instead of tick-tock? A flood watch.

What's the best time to tee off? In a driving rainstorm.

Where do pilots keep their money? In air pockets.

What did the snowman say to the beautiful snowgirl?
Do you believe in love at frost sight?

Is that natural snow on the ski slope? No, this resort
uses snow fakes.

My car engine sounds cold. Maybe you have it in frost
gear.

Seven days of shoveling snow makes one weak!

This rainstorm is ruining my topsoil. Blame it on mud-
der nature!

What do you get if you cross a sheep with a rainstorm?
A wet blanket.

What's the best weather for ducks? Fowl weather.

What do bees sing when it's raining? "I'm stinging in
the rain. . . ."

Will you stay up and watch the eclipse tonight? What
channel is it on?

Two fleas came out of a movie and one said to the other, "It's raining. Do you want to walk home or do you want to take a dog?"

Why did you cut a hole in your umbrella? Because I wanted to see if it stopped raining.

Mother: "Don't be selfish. Let your brother use the sled half the time." Daughter: "I do. I use it going down and he uses it going up."

What goes up when the rain comes down? Umbrellas.

Sign on beehive: WINTER IS COMING SO GET BUZZY.

How can you tell if a plane is about to hit turbulence? The flight attendant is serving coffee.

What happens when skiers get old? They go downhill.

What sport do you learn in the fall? Skating!

What's the hardest thing about learning to skate? The ice.

What's the difference between a person with a cold and a strong wind? One blows a sneeze and the other blows a breeze.

What's the difference between a horse and the weather?
One is reined up and the other rains down.

Why did the cross-country skier wear only one boot?
He heard there was one foot of snow.

What's the definition of a Seattle optimist? A guy with
a sun visor on his rain hat!

First pig: "I never sausage heat." Second Pig: "Yes, I'm
bacon."

What do you call a dog sled? A polar coaster.

Why is slippery ice like music? If you don't C sharp
you'll B flat.

What happened to the cow that was lifted into the air
by the tornado? It was an udder disaster!

Two meteorologists each broke an arm and a leg in an
accident and called from the hospital about their four
casts.

What did the homeowner say in the fall? Leaf me alone!

What do you get when you cross Teamsters with morning mist? Union dews.

Which part of the highway freezes first? The cold shoulder.

Teacher: What is the formula for a raindrop? Student: H-I-J-K-L-M-N-O. Teacher: That's not correct! Student: But you taught us that it was H to O!

Why is a hurricane season like Christmas? Because there's a good chance you could have a tree in your house!

Why did the obtuse angle go to the beach? Because it was over 90°!

What is the roofer doing out in the storm? He's shinglin' in the rain.

Will it rain tomorrow? Moist definitely!

What did Sherlock Holmes say about the overcast? "It's elementary my dear what sun?"

Iditarod: Mush transit.

Cold St. Patrick's Day: Erin go brrr!

Day Daay Daaaaaay . . . it's spring and the days are getting longer!

March Nor'easter: A brute spring scene.

Beach overexposure: Bakin' and aches.

Sunbathing: Oh tan and balm.

July has 31 days . . . summer hot, summer not.

Tourists and birds are heading south for the winter. They're brrrs of a feather.

Hurricane: A tree totaler.

Ozone alert: A bad air day.

Mostly sunny: When "I can't get no stratus action."

Low visibility: An act of fog.

Condensation: Much ado about dew.

Olympic forecast: Fair to medaling.

How can I see the sunrise? You'll see the glint eastward.

What's the wind forecast overnight? We'll have a night-in-gale.

An Army officer went to the Meteorological Office and told the meteorologist that a general was coming to the base and they wanted to hold an outdoor function in his honor. He wanted a forecast, and so the meteorologist studied his charts and satellite pictures. After a few minutes of thought, he told the officer with confidence, "The weather that day will be fine. You can go ahead with the party." The officer was very pleased and profusely thanked the forecaster. As an afterthought, he invited the meteorologist to attend the function. The meteorologist thought for a minute, and said, "Sure, I'd be delighted to come . . . if it doesn't rain." (S. Raghavan, India)

What's the King's favorite weather? Hail! (Ron Gird)

A husband and wife are home watching TV when the phone rings. The husband picks it up and listens for a moment. Then he says, "How should I know? Why don't you call the Weather Service?" Who was that, his wife asks. The husband says, "I don't know. . . . Some guy wanted to know if the coast was clear." (Nancy Dignon)

The First Rule of Forecasting:
Don't . . . it's too dangerous!
Corollary to the First Rule:
Don't . . . unless you have to because you're actually getting paid to! (Paul McCrone)

Why did the TV consultant say the meteorologist lost the audience during the weathercast? He spent too much time on the boreometer. (Matt Meister)

If my forecast didn't verify, it's obviously because the weather was wrong! (John Hathorn)

God sees the synoptic weather pattern from Heaven; you have a thermometer. Before you put out your forecast, don't you think you might ask God's opinion? (John Hathorn)

Good forecasts come from experience. Unfortunately, that experience usually comes from bad forecasts. (John Hathorn)

There are three simple rules for making good forecasts. Unfortunately, no one knows what they are. (John Hathorn)

In forecasting, you start with a bag full of luck and an empty bag of experience. The trick is to fill the bag of experience before you empty the bag of luck. (John Hathorn)

A computer has no stake in getting its weather forecast right. After all, it won't be taking its family to the beach this weekend! (John Hathorn)

Ever wonder why weather forecasters don't announce the day of *their picnic* to the public? (John Hathorn)

Baseball games have to be canceled in bad weather if the hailstones are three inches. How would they know which one is the ball? (John Hathorn)

If rainwater is so pure, where do the germs that give you the cold come from? (John Hathorn)

What did the meteorologist forecast for the rap concert? A Lil Wayne. (Eric Sorensen)

What did the cloud say to the sun? You're hot! (Holly Cartier)

Did you hear about the man who didn't know why the sun rises? He sat around and thought about it all night ... until it finally "dawned" on him. (Christy Henderson)

A child comes home from school after learning about clouds. There are many wispy ice clouds in the sky today, and the child is interested in them. He chases after his busy mom, who is on the phone. He finally catches up to her and says, Why so cirrus mommy? (Alexander Harrison)

Seattle Weather: "It tried to rain . . . but it mist."
(Attributed to Nancy Knight by Peggy Lemone)

My adiabat has lapsed
And my entropy is high
I've got rigor mortis
And I think I'm gonna die! (Peggy Lemone)

What did the updraft say to the moist parcel? If you follow me, I'll give you the condensed version! (Attributed
to Ben Green by Peggy Lemone)

Knock-knock Jokes

(Knock it off!)

Knock, knock.
Who's there?
Augusta.
Augusta who?
Augusta wind blew my hat off.

Knock, knock.
Who's there?
Wendy.
Wendy Who?
Wendy today, light breezes tomorrow.

Knock, knock.
Who's there?

Accordion.
Accordion who?
Accordion to the paper, it's going to rain today.

Knock, knock.
Who's there?
Lettuce.
Lettuce who?
Lettuce in, it's cold out here.

Knock, knock.
Who's there?
Turnip.
Turnip who?
Turnip the heat, it's cold in here.

Knock, knock.
Who's there?
Scold.
Scold who?
It's scold today, so let me in.

Knock, knock.
Who's there?
Atlas.
Atlas who?
Atlas the sun came out.

Knock, knock.
Who's there?
Gwen.
Gwen who?
Gwen it rains it pours.

Knock, knock.
Who's there?
Sonny and Cher.
Sonny and Cher who?
Sonny and Cher to cloud up later in the day.

Knock, knock.
Who's there?
Ivan.
Ivan who?
Ivan to come in, it's cold out here.

Knock, knock.
Who's there?
Armageddon.
Armageddon who?
Armageddon tired of this rainy weather.

Broadsides

All cartoons courtesy Jeff Bacon.

How forecasts are made

Weather terms to avoid

© Jeff Bacon of Broadside.net

In the spirit of process improvement, the Captain invites the
weatherman to the bridge to verify his forecast.

Earthquakes

(A whole lot of shakin' going on!)

Earthquake alert: That's groundbreaking news.

In California, you need no-fault insurance.

L.A. is the only place where you get turbulence when you get off the plane.

L.A. is a place where you get whitecaps in your coffee cup.

A lot of bric-a-brac was bric-a-broke.

There's going to be a big shakeup in the earthquake research laboratory.

What did one crack in the earth say to the other?
"Don't look at me, it's not my fault"

Did you hear about the wealthy philanthropist who left all his money to earthquake research? He was generous to a fault.

What did one volcano say to the other? "I lava you."

Lava come back to me!

It's Mother Nature making an ash of herself.

Magma come loudly.

Alaska

(Weather so crazy, it's the only state
with its own chapter!)

In Alaska a one-day sale lasts six months.

You wake up with arctic circles under your eyes.

Alaska: Where they have northern lights instead of traffic lights.

Alaska is a place where you can work for six months and call it a day.

What do you get when you sit on an iceberg? Polaroids.

Poster in an Anchorage travel agency: "Visit Alaska this summer. Three days, No Nights."

A man was on trial in Alaska. The prosecutor asked him, "Where were you on the night of October through May?"

An appliance salesman was trying to sell an air conditioner to an Alaskan. The Alaskan says, "Are you kidding? It's 50° below zero here." "Sure," says the salesman, "but what if it goes up to zero?"

Technology: Something that permits Alaskans to burn oil to produce electricity to run refrigerators to keep food cold.

Mushroom: The space between dog sleds.

Igloo: A domicicle.

Polar coordinates: Fashionable clothing worn by Alaskans.

Miscellaneous

(This book's least creatively
named chapter!)

A sky writer made an error. He was so upset that he flew
behind a cloud and wrote a dirty word.

The trouble with the morning is that it always comes
much too early.

We have the slowest traffic here. The cars even get
passed by the weather.

Bright and early? Personally I'd prefer it dim and later.

You have a better chance of being hit by lightning than
winning the lottery . . . but if you're hit by lightning you
can't buy a Ferrari.

If I won the lottery, the first thing I would do is go home and turn up my thermostat.

I love my job, but if I won the lottery I'd walk out right in the middle of my seven-day forecast.

What lies at the bottom of the ocean and shakes? A nervous wreck.

What do you get when you leave your calculator outside overnight? Dewey decimals.

Why did Dr. Jekyll go to the beach? He wanted to tan his Hyde.

I grew up in a dull neighborhood . . . even the rainbows were black and white.

They tried to put a heater in a kayak but it sank. You can't have your kayak and heat it too!

Oceanographers are deep thinkers.

It was the greatest disaster of 1996 . . . only exceeded by the introduction of the Macarena!

The weather is seasonable . . . that means there's salt, pepper, and paprika in the air!

Arizona had eight inches of rain . . . but it was a dry rain!

The meteorologist was nervous at his wedding. Instead of "I do," he gave the time and temperature.

We spend billions of dollars on satellites, radars, and sophisticated weather equipment. And what do we rely on? A window!

Window: An amazing observing system that permits you to observe the weather through solid walls and is hardly ever used by most meteorologists!

Norwegian meteorologist: A fjordcaster.

Oceanographer: A scientist with a deeploma.

Oceanography student: An aquademic.

Climatologist: A person who labels a region that's bitter cold in winter and sizzling hot in summer a temperate zone.

Meteorologist with a PhD: A person who has so many degrees he or she could start a heat wave.

I was a born meteorologist. The first words in my vocabulary were *possibly*, *probably*, and *unusual*!

I'm so old, when I was a kid, the rainbows were only in black and white!

A top secret government study indicates that we wouldn't be any worse off if economists predicted the weather and meteorologists predicted the economy.

Retirement is when you count the money you saved for a rainy day and there isn't enough now to handle a drizzle.

I heard that because of budget cuts at the National Weather Service, we're going to have weather only six days a week.

A man was interviewed for a job with the National Weather Service. A few days later he received this reply: "Your answers to our questions were vague, misleading, and indecisive. Congratulations, you're hired!"

A bank is a place where they lend you an umbrella in fair weather and ask for it back when it starts to rain.

If you know the difference between El Niño and La Niña, you've been watching too much weather on TV!

Did you know it was 10° at Little Bighorn today? That made it the first Frozen Custer Stand!

Today I was planning to go to the annual winter outdoor sports and wilderness survival show, but decided not to go out in this weather!

I know this TV meteorologist with a big ego. If anyone says, "Good evening" or "Good morning," he says "Thank you!"

If a tree falls in the forest and there's no one there to hear it, who fills out the environmental impact statement?

I've made a decision. I'm leaving my brain to medical science and my arthritic knee to the National Weather Service!

Three retirees were playing golf and one said, "Windy isn't it?" The second said, "No it's Thursday." The third replied, "So am I. Let's have a beer!"

There's plenty of places to stay in the desert. I once stayed at a Bedouin breakfast.

I saw my doctor at the gym, and I think he's getting a little chubby. He's a meaty urologist.

Barometer: An amazing meteorological device that can be used to confirm the onset of a storm if its reading is combined with torrential rain and powerful winds.

A man owned a lot of sheep and he wanted to take them over a river that was covered with ice. But the woman who owned the river said, "No!" So he promised to marry her and that's how he pulled the wool over her ice.

Wether: That's a bad spell of weather.

Cyclone: An exact duplicate of Cy.

Dewey decimals: What you get if you leave your calculator outside overnight.

International Date Line: "Hi, do you come here often?"

Kelvin Klein: Designer of thermal underwear.

Polyunsaturated: A parrot in a raincoat.

Vidal Monsoon: Meteorologist and international expert on correcting bad hair days.

The delegates to the last Global Warming Conference/ Save the Planet Summit had 12,000 limousines and 140 private jets.

Did you know that the weather radar was invented by a Russian meteorologist? Doppler Zhivago.

It's not easy being the child of a meteorologist. When she was in elementary school, the kids used to taunt her by yelling, "YOUR FATHER IS A LOW PRESSURE AREA!!"

The king banned all hunting, and soon the country was overrun with wild animals. When the people couldn't stand it anymore, they rebelled and toppled the monarchy . . . making it the first time a reign was called on account of game!

We just got a new designer radar. It shows thunderstorms in puce, snow in mauve, and drizzle in taupe.

In the olden days, people were much smaller than they are today. Knights rode on large dogs when they couldn't get horses. One dark and stormy night, a squire tried to purchase a large dog for his master, but unfortunately the shopkeeper could offer only an undersized, mangy mutt. The squire said, "I wouldn't send a knight out on a dog like this!"

Two fortune tellers were looking out the window at the snow coming down, and one said to the other, "Doesn't this remind you of the blizzard of 2025?"

Doldrums: An oppressive, sultry climatic zone bounded by lines of lassitude and loungitude.

There were two roosters outside when suddenly it started to rain. One made it to the henhouse . . . and the other made a duck under the porch.

Climate: What you do to a mountain.

Weather radio: The precipistation.

Thermostat: A device that keeps the house too cold for one person and too hot for another.

Warm front: What you get when you stand too close to the fireplace.

Weather: A condition of the atmosphere that permits people with absolutely nothing in common to start a conversation.

They had very heavy rain in New Orleans yesterday. It was the start of Muddy Gras.

I'm reading a book about an antigravity device and I can't put it down.

Earth's pull: Grabity.

California: A place where there's turbulence when you get OFF the plane.

Altimetry: The science of measuring height using an aneroid barometer, which attracts very few researchers because they can't stand the pressure.

Why did the rapper need an umbrella? Fo' Drizzle! (Mark Elliot)

Weather Observers

(If you see something . . . say something!)

Everyone talks about the weather, but how many people go out and buy barometers?

Weather observers are people whose business is always looking up.

Observers are people with their feet on the ground and their heads in the clouds.

Rain and snow make their day.

They're observing what nature is serving.

The people who never met a storm they didn't like.

The people who talk up a storm.

Rain or shine, they like it fine.

The people who have inside information on outside matters.

A really diligent observer called in to say it was dark.

The people who like to shoot the breeze.

The people who take a bow when they hear thunder.

A rough crowd: The only people who boo nice weather.

Observers are people who love snow. Obviously, they're flaky.

The Moon

(When it hits your eyes like a big
pizza pie . . . that's amore!)

We have a new moon tonight. I wonder what was wrong with the old one.

I wouldn't go to the moon if it was the last place on Earth!

During the eclipse, the moon will be just a shadow of its former self.

Tonight you could go out on a moonlight date and come home in a fog.

Nerdiest place in outer space? The dork side of the moon.

The moon is almost full. We'll empty it in the morning.

If you have skin that burns easily, don't forget to put on your moonscreen.

The moon is full. The question is, What is it full of?

If you're thinking about taking a trip to the moon forget it . . . the moon is full.

On the moon, it's 260° Farenheit during the day and 270° below zero at night. Sort of makes it hard to pack light!

I couldn't watch the eclipse so I taped it. Don't tell me what happened!

You know you're old when it hurts to stare at the moon.

They found water on the moon . . . but you have to ask for it.

Two guys were leaving a restaurant during last night's lunar eclipse. One guy said to the other, "Is that Earth's shadow moving across the moon?" The other replied, "I really don't know. I don't live around here."

Moonlighting: The sun's other job.

Eclipse is what a gardener does to a hedge.

Only a small sliver of the moon was visible in the sky, and I was wondering if it was waxing or waning? I decided to check the weather page of the newspaper where things like this are spelled out in detail. This didn't really settle my question. The overnight forecast called for a 60% chance of wane.

Two people were dining in the first restaurant on the moon. After looking around, one said, "I don't like this place, it lacks atmosphere."

Air Pollution

(Dirty weather jokes!)

It'll be a while until they make a car that's totally pollution free. . . . That's known as emission impossible.

For years engineers have been trying to remove pollution from automobile emissions, but the work has been too exhausting.

If gasoline prices keep going up it will make air pollution out of reach for the average person.

Does this make sense? A 10,000-car motorcade to protest air pollution!

A good remedy for acid rain is a Tums umbrella.

Do you think the ozone layer over Switzerland has a lot of little holes?

To solve the air pollution problem the government has to stick its business into other people's noses.

Sure I worry about smog, ozone, and acid rain ... but how do you tell your kid not to stand too close to the air?

All of our waters are filthy. Recently it was reported they discovered water under Long Island Sound.

I went down to the beach and put a shell to my ear and heard coughing!

Our water supply has to be polluted. I walked by the harbor and heard fish wheezing.

If they don't start cleaning up our beaches, one day the tide will go out and refuse to come back in!

They claimed they cleaned up the beach, but I think they left oilier than expected.

Mary had a little lamb as dirty as a hog. I asked her how it got that way. .... She answered simply, "Smog."

Air pollution: The lung and short of it.

What happens when radioactive waste is dumped in the ocean? You get nuclear fishin'!

With today's air quality, I'm up to two packs a day . . . and I don't even smoke!

We need to make people who pollute the air pay through the nose!

Acid rain: The forest's prime evil.

Environmental pollution: Assault of Earth.

Environmental pollution: The Grossest National Product.

Environmental pollution: Domain poisoning.

Smog: Mother Nature making haze while the sun shines.

Smog: A form of air pollution that lowers your lungevity.

Just a tip if you're planning a summer vacation: If you're going to an undeveloped country, don't drink the water. If you're going to visit a developed country, don't breathe the air.

Highway pollution: Auto Oxidents.

We have an Air Quality Management District, which gives you the impression that we manage the air . . . but the air does pretty much what it wants to.

What do you see in Southern California on a clear day? UCLA!

This air is so bad . . . parents are telling their kids not to stand too close to the air.

As to the question whether it's safe to breathe this air? I say yes, but don't inhale!

Yesterday I went outside in New York to get a breath of fresh air . . . and I wound up in the Adirondacks!

I can't stand this polluted air! I happen to be a chain breather.

I had to step inside for a breath of fresh air.

The air was so bad . . . I stepped outside to take a deep breath, and I chipped a tooth!

I stepped outside to get a bite of fresh air.

Remember when the expression "There's something in the air" was just a figure of speech?

You no longer use the word pollution. . . . Today it's "artificially flavored air."

The pollution is so bad . . . I just ate a bagel and could taste the hole!

Space

Those space projects make me wonder if there is intelligent life out there or if they're just like us.

I believe there's intelligent life in outer space. In fact, they're so intelligent they won't have anything to do with us!

Knock knock.
Who's there?
Astronaut.
Astronaut who?
Astronaut what your country can do for you, but what you can do for your country!

If space scientists are so smart, how come they measure distance in light beers?

One astronaut is giving NASA a hard time. He says the weather is nice in space and he wants to ride with the top down!

Astronauts are the only people who become heroes when they're down and out.

Orbiting junk has become a problem to spacecraft. NASA's solution is to send a giant Electrolux into orbit, but they don't know if a vacuum can work in a vacuum.

They're thinking of sending cattle into orbit. It will be the herd that's shot around the world!

Mars has a very unfriendly environment. Temperatures are below zero and it has a poisonous atmosphere, which means they're having a better summer than us!

The idea for this space project came in second at a science fair . . . it won a constellation prize.

What kind of shampoo do you use in meteor showers? Head and boulders.

If the universe keeps expanding, how come the sky never gets any bigger?

The probe to Mars takes off today and will land back on earth July 1 if there's no holiday traffic.

The winds on Saturn over 700 mph. Astronauts would have to use a lot of hairspray!

He's the kind of guy who would yell "ASTEROID!" in a crowded planetarium!

I don't know about the space program. Outer space is the last place on Earth I'd want to go.

Thanks to the invention of the telescope, objects that are 100 trillion miles away appear to be only 50 trillion miles away.

When do astronauts eat? When they go out to launch.

NASA scientists made a remarkable discovery: The rings around Saturn are composed entirely of lost airline luggage.

Last night we went to the planetarium. Not only was there a cast of thousands but everyone was a star!

Scientists announced they've made contact with space aliens whose planet is completely covered by one gigantic shopping center. Skeptical scientists didn't believe it initially but have now confirmed "that it's a mall world after all."

It was the little boy's first trip to the planetarium and he wanted to make a reservation for a rocket trip to the moon. "I'd like a ticket to the moon," he told the clerk. "Sorry young fellow," the clerk said, "all trips to the moon have been canceled." "Why is that?" the boy asked. "Well you see," said the clerk, "the moon is full."

Astronomer: A person who is star-craving mad.

Exorbitant: An expensive weather satellite that has fallen.

On the date of the equinox, the sun crosses directly over the equator, and it's said you can balance an egg on its end . . . the bacon strips are a lot harder, and forget about the toast!

Spring

(For when you have spring fever . . .
and it isn't even spring!)

April is the month when the green returns to the grass,
to the trees, and to the IRS.

It's hard to sleep with all this germinating going on.

April is a good month because the weather starts to get
warmer just about the same time the IRS takes the shirt
off your back.

April is a month that comes in like a lamb . . . because
you may still need to wear wool!

Time for the kids' annual spring ritual . . . wearing much
lighter clothes when they stay inside to play video games.

Best thing to drink during a marathon? Running water!

March is a month with a credibility gap. When it snows, it's wet, and when it rains, it's freezing!

March gives us a rerun of the weather of the past few months and a preview of what's to come.

March is not one of my favorite months. It's hard to believe the calendar would rush us through February just for this!

The principal function of March is to use up all the rotten weather that wouldn't fit into February.

March comes in like a lion, goes out like a lamb . . . and usually gets your goat!

March comes in like a lion and goes out like a lamb . . . kind of like most of us when we ask for a raise!

I don't object to March coming in like a lion, but I resent it when it hangs around like a polar bear!

If you expect March to act like a lamb, you're a dyed-in-the-wool optimist!

Will March go out like a lamb or will it be ba-a-a-ad?

When March acts like a lamb, it's usually pulling the wool over your eyes.

This is the time of year you look forward to two kinds of March madness . . . one is about basketball and the other is the delusion that the weather will stay nice for two days in a row!

March: It's a month you're happy to see arrive and even happier to see depart.

The days are getting longer, which is good because I always thought 24 hours were never enough.

Spring is when you turn on your heat the day after you turn it off for the season.

This week I'm running a special deal. I'm clearing out all merchandise to get rid of all my "how cold was it?" jokes. Everything must go!

It's spring. Now all the people who drive like idiots in the snow can drive like idiots in the rain.

The start of spring divides people into two categories: those who are happy and those who ski.

Spring: When birds sing, flowers bloom, and mud slides.

How cold is it? Cold enough to chill a mockingbird!

Enjoy these brief few days between "hot enough for you?" and "cold enough for you?"

It's spring and a young man's fancy turns to golf . . . just like the rest of the year!

It's spring when both you and your car get put on a salt-free diet.

Spring . . . it's time to plant those radishes, those onions, those golf tees.

I noticed even the birds are cranky. They had to return from their vacation.

Spring is when the animals come out of hibernation and skiers come out of traction.

It's spring. And if your fall leaves haven't blown onto your neighbor's lawn by now, it's not going to happen.

Isn't this fantastic weather? I think I just saw the first penguin of spring.

Sometimes there's a very big difference between the first day of spring and the first spring day.

It's the time of year when the season seldom gets together with the weather.

It's when farmers and golfers both do their plowing!

It's when there's an urge to dig in the dirt. Some do it with a garden hoe and others do it with a golf club.

This year I built three snowmen in my yard. It's not that I enjoyed doing it. . . . I'm planning to take them off as dependents!

I just saw an interesting sight . . . a bunch of accountants migrating back to the city for the tax season.

It's when fishermen get that faraway lake in their eyes.

It's when golfers get that fairway look in their eyes.

It's spring at last . . . the birds are singing, flowers blooming, Mets sinking, Knicks choking. . . .

It's a nice time of year. It's when *damn* and *snow* become two separate words again.

This is when we go from snow blowers to nose blowers.

It's the start of the allergy season and I'll sneeze to that!

May is such a nice month . . . you would think it wouldn't have any Mondays.

Have you heard the song "Blowin' in the Wind?" I like Peter pollen Mary's version!

The pollen count will be increasing. . . . I thought I'd tell you that just in case you want to make your own honey.

The pollen count is high so all unnecessary breathing should be curtailed.

Sneezings greetings!

No one ever talks about the good side of the allergy season. Do you realize that if it wasn't for coughing and sneezing, some people wouldn't get any exercise at all?

The month of May isn't very pleasin' to those who spend it sneasin'!

You know you're having a bad allergy season when you wake up and hear birds coughing and sneezing!

May is Mother Nature's way of apologizing for February.

May . . . a month that's outdoorable.

I can't understand why May is so popular. Where was it in February when we needed it?

It's June . . . the weather is great and thousands of cheerful, smiling, and shining faces are looking forward to their vacation. They're known as teachers!

It's spring and the shortest distance between two points is under repair.

June is busting out all over . . . I don't know what that means but it sounds unpleasant!

June . . . a wonderful month with much ado about "I do."

Scientists are trying to solve one of life's great mysteries . . . why a lawn mower that's been resting all winter in your garage refuses to work in the spring.

My neighbor nearly went crazy trying to teach his dogwood to fetch.

Good news! My dogwood had puppies.

I saw an interesting spring sight . . . my dogwood was chasing a pussy willow.

Spring is Mother Nature's way of saying, "Let's Party!"

Spring—my favorite winter vacation!

I just saw my 165th robin of spring. Most people stop counting at the first.

People are excited. Everything is turning green. Big deal ... I have things in my refrigerator that do that all the time!

The calendar says it's spring, but you have to allow several weeks for delivery!

Spring is a divine season to the poet, but it's just sleet and rain to the meteorologist.

Spring is the time of year when everything begins to blossom and grow ... except what you planted!

Spring is when the furnace repair companies head for the southern beaches and the air-conditioning repair companies return.

I have only two bare spots to reseed this year—the front yard and the back yard.

Spring: When the sun smiles and the earth greens back.

Today is the start of spring. It's a root awakening!

Today is the start of spring. Seasons greenings!

Change of seasons: Meadowmorphosis!

Do you know when spring really arrives? First it gets warm and everyone plants their garden. Then it gets cold and everything freezes. Next it gets warm again and everyone plants their garden again . . . and THAT'S spring!

I was a little late today . . . I forgot to set my watch ahead. But I only get to use that excuse once a year.

This has been a bad week. A few days ago I turned my clock ahead and lost one hour. Today I watched the Academy Awards and lost four hours!

Moving the clock ahead one hour seems to work so well. Why don't we move the clock ahead 24 hours? Then all the rain we get on Sunday will fall on Monday instead!

Why can't we have a Weight Saving Time and take numbers off our scales? Call your Congress representative!

If we can move our clocks ahead one hour, why can't we move our clocks ahead one month . . . then it will be April!

I'm still on daylight saving time, so I'm one hour ahead. If my forecast doesn't make any sense, you'll understand it an hour from now.

They say you gain an hour. No way! By the time I adjust my computer, my clocks at home, and in my car, I've lost *two* hours!

Daylight saving time gives golfers an extra hour to look for their ball!

Daylight saving time is really confusing. I stayed up all night to see what time the sun rises . . . and then it dawned on me!

The way things are going when you give up an hour in March, you're never sure you're going to get it back in November.

During daylight saving time, do they change the name of 7-11 stores to 8-12?

Spring ahead . . . fall back . . . it's like me trying to get out of bed in the morning!

It's the end of daylight saving time and you gain an hour. It takes me more than an hour to put the wall clock back on that little nail!

The dirty little secret: Not a single second of daylight is saved by this ploy!

Why do we have to lose an hour on a Sunday? Why can't we lose it on a Friday?

I love the end of daylight saving time. . . . I get paid one hour earlier!

I've been going crazy trying to figure out what to do with the hour I gained today.

Now who looks like a genius for not setting the clock in my car ahead last March!

Personally I'm against daylight saving time. At my age giving away even one hour isn't so smart!

It's interesting that the sun always goes down at sunset. No one knows how it adjusts to the change in time.

I could never understand why our government goes to such great lengths to save daylight. Let's be honest . . . don't we have a lot more fun at night?

I like switching back to standard time. It gives me a feeling of power in that I can make the sun come up one hour earlier!

Spring ahead . . . fall back . . . but enough about my career path!

I'm not going to lose any sleep over losing an hour's sleep!

Don't bother turning your clock ahead. It's easier to just show up for everything an hour late.

St. Patrick's Day: It's the one day a year I turn my forecast over to leprechauns!

The storm hit everywhere but the garden . . . it's the luck of the iris.

He was named Pat because he was born on St. Patrick's Day. It's an honor to be named after a holiday, but you have to feel sorry for his sister Groundhog!

This is an IRS rain. You can't get through it without getting soaked.

April 15 . . . it's the original spring cleaning!

Easter will be partly bunny!

Our forecast is for a nice day for Mother's Day. We'll see Mom-shine!

For Father's Day I received a barometer made in Japan
... now I can tell if it's raining in Tokyo.

Wearing those black graduation gowns on a hot June
day is ideal preparation for the real world ... it gets
them used to sweating.

Graduation: It was a brainy day!

It was graduation and those people received so many
degrees they could start their own heat wave!

Meteorology degree: Magna cumulus laude.

An even higher achievement on a meteorology degree:
Sunny cum laude!

To save the environment, my forecast was constructed
from recycled weather.

I'm Earth-friendly ... I decided all my forecasts are
going to be recycled.

Ever since Earth Day the weather has been treating us
like dirt!

Yesterday, we celebrated Earth Day by throwing mud at
each other.

For Earth Day all my forecasts will be biodegradable.

Today is Earth Day, so I'm giving tonight's seven-day forecast using very little energy.

Every time you breathe you're spewing carbon dioxide into the air . . . so I'm encouraging you to stop breathing for Earth Day!

Earth Day celebration: Ground Hug Day!

Earth Day? We named our planet after dirt?

Tomorrow is Earth Day. It's so hard to shop for the earth, so I usually just send a card.

April showers bring May flowers, flooded basements, and kids cooped up in the house for hours.

People who like April showers are all wet!

I respect tradition . . . so in April I always take a shower.

This seems like a rotten way to bring May flowers!

Do you believe that for every drop of rain that falls a flower grows? Well, it looks like we're going to be up to our noses in petunias!

April showers bring May flowers. However, a little seed-ing, weeding, and fertilizer help.

June is rusting out all over!

June doesn't always bust out all over . . . sometimes it just leaks in.

I can tell it's spring today. I saw a robin rust breast!

Heavy rain, high winds . . . it looks like Mother Nature has started her annual spring cleaning!

It's so cold in New York, the Mets will be looking for the home ice advantage.

If there's anything I hate it's winter going into extra innings!

Today it's going to be a rainy game . . . the Damp Yan-kees versus the Wet Socks.

You can tell summer is approaching . . . the rain is get-ting warmer.

People are bringing out their rain-making equipment otherwise known as barbecues.

Spring

I love barbecues in the rain ... it wets my appetite!

Neither rain, nor snow, nor sleet ... I'm not talking about postmen ... I'm talking about golfers!

I know why so many poets love spring ... you can rhyme so many things with rain!

I know it's spring. You know it's spring. But does Old Man Winter know it's spring?

Summer

(Never let them see you sweat!)

Today is the first day of summer. Seems like just yesterday, it was spring!

Ever wonder where Zambonis go in summer? Do they migrate?

My goal this summer is to develop as deep and rich a tan as my lawn.

No matter how hot it gets . . . don't sweat it!

Summer is when you pay a kid $15 an hour to mow your lawn so you can get to the gym to work out!

Summer is when you don't do all the chores around the house you wanted to do all spring.

Summer: It's that long, uncomfortable season between a pleasant week in the spring and a pleasant week in the fall!

It's the season that bugs us.

It's bug season . . . of course, I bug people in every season!

It's August . . . it's a sad time when it's too late to start a garden or slim down for the summer.

Summer is starting to make me sweat. I just got my electric bill!

I just listened to the sound of the ocean . . . and it sounded just like the inside of a seashell!

Ocean shore: Where buoy meets gull.

Summer fun is like winter fun . . . but with sunburn instead of frostbite.

Nude beach: Where there's nothing like a day in the sun to put color in your cheeks.

Beach overexposure: The wages of sun.

I have a new product . . . low-carb sunblock.

People who stay out in the sun too long are just basking for trouble.

I use SPF 75 . . . it's so strong it not only blocks the sun but makes it rain!

I was grilling outside today . . . I forgot my sun screen!

Tangent: A guy just back from the beach.

I'm an old-timer . . . I can remember when sunshine, bacon, and eggs were good for you.

People are baking on the beach. I could tell one woman was done because her belly button popped up.

The beach was so crowded there was no room for the tide to come in.

Boats are for people who can't get sunburned enough on land.

I hate boats. I get seasick when I give the marine forecast!

Looking at these people laying around on the beach it's hard to believe the sun is our primary source of energy!

I love the beach ... a gentle breeze, bright sun, warm sand, and the sound of surfers crashing against the rocks.

Summer is exciting. Surfers are looking for the perfect wave and sharks are looking for the perfect surfer.

The weather down at the beach was getting wave reviews.

Sunbather: A person who is well red.

Sunbather: A baked being.

Sailing lesson: On the jib training.

If 70% of Earth is covered by water, how come it takes so darn long to get to the beach?

Ah the smells of summer ... chlorine, lighter fluid, bug spray, calamine lotion ...

You know what's scary about summer traffic? There's just as much traffic on the other side of the road trying desperately to get away from the place you're heading!

You have to wonder . . . do professional fishermen take a week off to go accounting?

I received a compliment today. I was told I had a nice pale!

On August 1, 1774, Joseph Priestly discovered oxygen. Before that date, people would just breathe whatever they could get their hands on!

The longest day of the year . . . it's not fair! We've waited for summer for so long and now they start taking time off of daylight!

It's summer. People put up screens against bugs, air-condition their homes, and then go outside for a barbecue!

It's a hot Independence Day . . . so go fourth and multi-fry!

People tell me I look healthy, but under this tan you should see how pale I am!

I'm putting on weight so I can get a bigger tan.

Studies have shown that the hotter it gets the more romantic people become. Isn't it possible that the more romantic people become the hotter it gets?

Capital punishment: Spending the summer in Washington, D.C.!

I finally figured out our weather pattern: Hot in August and cold in January, but not necessarily.

This is the time of year you look at your bill for air-conditioning... and lose your cool!

There are two reasons to cancel mowing your lawn: (1) rain and (2) shine.

The weather during the month of August shouldn't happen to a dog!

Sunbathing: A fry in the ointment.

Dehumidifier: A device to get rid of de humidity.

What did the pig say in yesterday's hot weather? "Whew, I'm bakin'!"

Undertow: What surfers get instead of athlete's foot.

Summer is when you need iced drinks to cool you off from the warmth of the sweater you're wearing to ease the chill of the air conditioner you have on because it's summer.

I'm glad summer's almost over . . . I'm tired of using a pot holder to open my car door.

This is the time of year we go from thunderstorm warnings to two-minute warnings.

Summer just zipped by. You'd think it would have slowed down in all this heat!

An ant came out of its hole today and saw its shadow, which means we'll have six more weeks of picnic weather.

It's the time of year the days get shorter . . . and the faces get longer!

This is the time of year most folks would like to have as good a tan as their lawn.

Summer is interesting . . . you look forward to it all year, complain when it's here, and are sorry when it's gone!

I went down to the beach and put a shell to my ear. It said, "Summer's over."

If there's anything to reincarnation, I'd like to come back as an air-conditioning unit. You work three or four months and then you're off the rest of the year!

If it was such a bad summer, why are we sorry it's over?

Summer is out and fall is in . . . and we go from swimming pools to football pools.

Labor day: A case of summer flew.

Autumnal equinox: When summer's gone in one fall swoop.

Air-conditioning is wonderful! You can breathe cool air day and night for only pennies a breath.

If you miss the good old days . . . turn off your AC!

Sign in window of barber shop: "HAIR CONDITIONED"

Air conditioner: A device that helps you keep cool until you get your electric bill.

How come there are fan clubs but no AC clubs?

I turned on the humidifier and the air conditioner . . . and it started raining in my kitchen!

Air conditioner: A device for controlling the comfort level for humid beings.

Air conditioner: A useful summer appliance whose chief virtue is that your neighbors can't borrow it.

May the force be with you and may the force be air-conditioned!

Why don't we call them the hot dog days of summer?

It's the dog days . . . except in Quebec where they're called "les jours de bow wow!"

In this weather you find out who your REAL friends are . . . they come and visit even if you don't have AC!

I spend a lot of time in the library during the summer . . . and every other place that has air-conditioning I don't have to pay for!

I'm trying to figure out which is colder . . . when you turn your AC up or when you turn your AC down?

It's confusing. In this heat you're advised to drink plenty of juice, and, on the other hand, the government says, "Use Less Juice."

Summer is when a person will go from an AC house to an AC car to an AC health club . . . to go into a sauna!

An old-timer is someone who raves about the good old days . . . while riding in an air-conditioned car in this heat!

I love AC . . . now I can have a cold any time of year!

Will all the people who have window air conditioners, please turn them around full blast? That should make it cooler outside!

I have a solution to global warming . . . global AC!

I've said this before and I'll say it again . . . a temperature of 93° is better than no temperature at all!

If you're headed for the beach, it won't be the heat but the humanity!

If you're camping out in this weather, the heat will be in tents.

What did I tell you? That warmer weather you wanted back in January finally arrived!

These are times when you understand what the meteorologist has in mind when he talks about mean temperature.

I hope you realize . . . the same people who are complaining about the heat are the same people who complained about the cold last January.

Because of today's heat, the movie tonight on TV has been changed from *Some Like It Hot* to *The Big Chill*.

If this weather is getting to you, it helps to be reminded that in 120 days you'll be complaining about your heating bills!

Barbecuing is a funny activity. . . . It's 100° and the humidity is 95%, so what do you say? Let's light a fire!

It's so hot, I'm issuing a severe hug warning. If you hug someone there's a 60% chance you'll stick together permanently!

I'm going to start using Centigrade. Doesn't 32° sound a lot cooler than 90°?

If you want to feel cooler, take the cable off the back of your TV and watch the snow.

There's a new Post Office motto: "Neither rain nor snow . . . but if it's this hot again tomorrow forget it! I'm going swimming."

It's the dog days of summer. Apparently "dog days" is another expression for horrible.

These dog days don't bother me as much as the mosquito nights.

It's called the dog days because people wear shorts . . . and you can peek-on-knees!

Looks like I'll be traveling this summer . . . dragging my chair in front of the AC.

To beat the heat, take a lesson from the clever Swiss. . . . Live in Switzerland!

I want you to know it's both the heat *and* the humidity!

You heard the expression "It's not the heat but the humidity?" No! Sometimes it's just the heat.

There's plenty of whewmidity.

It was so humid . . . I had to chew to breathe!

I want to take a poll. How many people out there are sweating in this hot and sticky weather? But PLEASE don't raise your hands!

You know it's humid when you have a rainbow in your living room.

Before I begin, can I please have a nice round of applause . . . just to stir up a breeze!

I played tennis today . . . it's not the heat, it's the stupidity!

For me it was a nice day . . . not one person reminded me about the humidity.

This is the time of the year when everything that's supposed to stick together comes apart, and everything that's supposed to come apart sticks together!

It's the time of year the chair you're sitting on gets up when you do.

I took the clothes I wore yesterday and threw them in the hamper . . . and the hamper threw them back!

If you exercise in this weather, it's a good idea to drink liquids . . . it's uncomfortable drinking solids!

The air is so thick . . . if you hit it with a stick, it rains for a couple of seconds.

It was so hot I saw a hitchhiker on the parkway holding up a sign: "AC only!"

What I wouldn't give to be able to see my breath again!

There is a cold front on the map, but I expect it to miss us by at least 13 states.

What did the pig say on a really hot day? "I never sausage heat."

It hit 102.5° today. That's not a temperature! That's a radio station!

Doesn't that blizzard we had last winter sound good right now?

It may be 90° to us, but that's 630 in dog degrees.

Because of the heat, the highway department is advising all motorists to drive in the shade only.

We had a temperature of 90°... but with the wind chill factor, it felt like 89°.

The heat caught the power company with its plants down.

It was so hot pitchers at the ballpark were deliberately throwing home run balls so they could be sent to the showers!

They say there's no humidity today . . . there's no humidity in my oven either, but I can still cook a turkey!

We're lucky it didn't snow. Wouldn't you hate to shovel snow in this heat?

My wife and I were having an argument, but it was so hot I stood next to her just to get the cold shoulder!

It was so hot the penguins at the aquarium were down to shirtsleeves and cummerbunds!

It was so hot . . . people on Wall Street didn't mind losing their shirt today!

It was so hot . . . people were actually driving with those cardboard sun shields in their windshields!

It was so hot . . . on Broadway they were doing *Fiddler on a Hot Tin Roof*!

It was so hot . . . the weather observer's thermometer read "continued on next thermometer!"

It was so hot . . . the only breeze in New York was from the Mets players striking out!

Don't go jogging in this weather . . . it's too hot to trot!

Forget about using sunblock . . . use barbecue sauce!

On a day like today you could work up a sweat taking a cold shower.

New summer replacement show on TV: "Sweating with the Stars."

If someone tells you it's hot enough to fry an egg on the sidewalk, don't let them cook breakfast for you!

After the last heat wave my electric bill read $236,000. It was a mistake, of course, and the power company said, "Just go ahead and pay it, and we'll take it off next month's bill."

Summer is easily my favorite season . . . except, of course, when it's actually summer!

Isobar: An air-conditioned saloon.

Tangent: A guy just back from the beach.

Boy is it humid out there! How humid is it? It's so humid, the dew point is over the don't point. . . . I don't want to dew anything! (Todd Glickman)

You're aware that the summer solstice was yesterday. It's the longest day of the year . . . in fact, it's still yesterday!

The Sun

(Jokes to brighten any day!)

If you get up at dawn to see the sunrise, you couldn't have picked a better time.

If you listen closely, you can hear the crack of dawn.

One of the good things about solar power is you can look up and see if you have any of it left.

Sundial: An old-timer.

Get away from that TV and watch the sunset. The special effects are spectacular, and it's free with no commercial interruptions.

Total solar eclipses scare the daylight out of me!

The sun is a celestial body whose light energy moves so fast it gets here much too early in the morning.

I used to be a vegetarian, but it had side effects. I always found myself leaning toward the sun.

The trouble with dawn is that it comes much too early in the day.

There's nothing prettier than a summer sunrise. But it would be even prettier if it came around noon!

My meteorology professor once told me: "It's always darkest at night."

Once again the sun is rising in the East. . . . I think I'm beginning to detect a pattern here.

In the Caribbean, if the moon passes in front of the sun they call it a total calypso of the sun.

We all know about the speed of light . . . but what about the speed of dark?

I'm a light eater. I start eating as soon as it's light.

She was an eccentric mother. She wanted her son to sit on her bread dough . . . because she enjoyed watching her son rise in the yeast.

Eclipse: What a gardener does to a hedge . . . or a barber does to hair.

God: "Whew! I just created a 24-hour cycle of alternating light and darkness for Earth. Angel: "What are you going to do now?" God: "Call it a day!"

Sun: The oldest settler in the West.

Sunrise: A spectacle that millions of people would get up to look at if they had to pay to see it.

Unusual Weather

(The weather we usually get!)

This unusual weather is more unusual than the usual unusual weather!

Unusual weather is a term used to describe the usual weather around here.

Too much rain causes floods. Not enough rain causes drought. Notice no one ever says that "we had just the right amount of rain."

The barometer has been rising and falling so fast this week I had to take a Dramamine before I could prepare my forecast!

Talk about a change in the weather . . . you can get a suntan right over your frostbite!

It's the first time I ever saw lawn mowers with snow tires!

This cold is affecting the local wineries . . . there's a small carafe advisory!

Some of the vineyards are changing their wine to Cold Duck.

I went to the beach to get some color and got it . . . blue!

What cold weather! It's the dog sled days of summer.

Fathers have been telling their children, "When I was young, the days were so cold that rain would actually freeze and coat the ground with this frozen white stuff!"

This weather is proof that Mother Nature is really a committee.

Mother Nature gives us these changes in the weather so people not interested in sports (or without kids) will have something to complain about.

What happened to warm weather? This is the green-house defect!

This unusually cold weather is Mother Nature's way of saying, "I'm really sick and tired of seeing you in shorts!"

Tall Tales

(And every one is the honest-
to-goodness truth!)

After volunteer parishioners finished painting the church exterior, a dark cloud appeared and the heavens opened up washing away the newly applied paint off the church. The preacher was in tears and the congregation was stunned. Just then there was a flash of lightning and a loud voice from the heavens rang out, "REPAINT! REPAINT! AND THIN NO MORE!"

A little kid who was growing up in the city and not familiar with the beauties of nature was on vacation in Niagara Falls. There, he saw his first rainbow. As he looked upon this gorgeous site, full of wonder, he said to his mother, "It's really beautiful, but what does it advertise?"

The film crew was on location in the desert. One day a shaman approached the director and said, "Rain tomorrow." The next day it rained. A few days later, he said to the director, "Storm tomorrow." The next day there was a hailstorm. "This shaman is incredible," said the director, "I have to hire him!" After several more successful predictions the wise man didn't show up for several weeks. Finally the director called for him. "I have to shoot a big scene tomorrow, and I'm depending on you," said the director. "What's the forecast for tomorrow?" The shaman shrugged his shoulders. "Don't know," he said. "Radio broke."

The little polar bear came home from school one day and asked his parents, "Am I really a polar bear?" The parents said, "Of course you are." A few days later he asked the same question and got the same answer. About a week later he asked, "Were your parents always polar bears?" His father asked why he wanted to know if he was really a polar bear. The baby bear said, "Because I'm always shivering!"

The man was lining up his putt on the eighth green when suddenly a woman dressed in a bridal gown came running toward him. "It's our wedding day! We're supposed to get married," she shouted. "How could you do this to me?" "Listen," the man said. "I told you only if it's raining . . . only if it's raining."

No matter how important you think you are or how many awards you've received, if you remember one thing, it will keep you humble: The turnout at your funeral will depend on the weather that day!

The family owned a small farm in Canada just yards away from the North Dakota border. For generations, their land had been the subject of a minor dispute between the United States and Canada. One day they received a letter stating that the Canadian government had reached an agreement with the politicians in Washington, D.C. They decided their farm is really in the United States. "This is wonderful," the mother said. "I don't think I could stand another one of those Canadian winters!"

An artist specialized in paintings of coastal towns ravaged by hurricanes and Nor'easters. At a gallery, a visitor studying his many paintings said to him, "It's a shame you always have such bad weather!"

Two cavemen sat together near a fire as lightning flashed and thunder boomed. One caveman said, "We didn't have weather like this before they invented the bow and arrow."

From a radio broadcast: This just in . . . more football scores: Boston 42, Buffalo 35. Philadelphia 55, New

York 47, Albany 48. . . . Oops, excuse me! I've been reading the latest weather observations.

On a cold winter morning, a TV meteorologist announced that the temperature was 12°. However, he said there was probably a wide variation in temperature throughout the area due to effects of the ocean. He asked the viewers if they would check their thermometers at home and e-mail the weather department so he could make an unofficial survey. One observer replied that he had checked his thermometer and his temperature was 98.6°.

On April 16, 1875, two weeks before the event, the North Dakota Derby was moved from Fargo to Louisville due to heavy snow and bitter cold. It was renamed the Kentucky Derby . . . and this move increased the attendance from 147 people to 135,000!

In 1938, the Syracuse University baseball team, playing on the road, became snowbound during a rare spring blizzard, trapping both teams in a 26-foot snowdrift. The Syracuse players lived on a bushel of oranges they had with them, hence the nickname "Orangemen," which was then used for all the school's athletes. Their opponents found the experience more trying, having to eat nothing but toothpaste. That team's school later became known as Colgate University.

The great drought in the Midwest during the 1930s, which resulted in millions of acres of farmland becoming useless and forcing hundreds of thousands of people to leave their homes, came to an end when it was discovered by two government scientists that you could make it rain by washing your car.

In 1943, a U.S. destroyer collided with an iceberg in the North Atlantic and was sunk. Five sailors managed to climb aboard the iceberg and survived. In the name of the U.S. Navy they commandeered their new ship of ice and named it the USS *Kelvinator*. They spent the next several years attacking German ships and were a vital cog in our taking control of the North Atlantic shipping lanes. They were too aggressive one day when they chased a German U-Boat near Bermuda. It remains today as the only ship in our naval history to be melted in action.

In Nebraska it was so hot that farmers had to feed their chickens cracked ice to keep them from laying hard-boiled eggs.

During the middle of a drought in Texas, some friends were discussing the weather when a farmer asked how the lack of rain was affecting them. "Well," said one man, "I've got fish in my pond six months old that ain't even learned to swim yet."

It had gotten so dry during a drought in Mississippi that catfish were coming directly up to homes to beg for water.

When I was a kid, it was so cold that when we tried to talk our words would freeze. So we tied them into bundles and carried them into the house. Then we would hang the words over the fireplace. When they thawed, we could speak to each other without saying anything.

Flavored snow that could be scooped up in cones and served was developed by a Kansas entrepreneur who went up in a crop duster and sprinkled the snow clouds with vanilla. His venture was stopped by the Public Health Department and his plan for fudge-coated hail had to be abandoned.

One day it was so hot outside, I took a hamburger patty out of the freezer and tossed it up in the air. When it came down it was cooked well done. You had to be careful not to toss it up too high or it would come back down burned.

I asked a farmer in Nebraska, "Does the wind always blow like this?" "No," he told me, "about half the time it backs around and blows the other way. In summertime, the wind blows so hard it causes the sun to set three hours earlier than it does in the winter."

An Iowa farmer said the sun was so intense, his cornfield exploded into fluffy white clouds of popcorn. His cattle thought it was snowing and froze solid on the spot.

Some advice for winter fishermen who fear their catch may be tainted with mercury: Hang the fish tails down, then wait for the temperature to drop to subzero. At that time the mercury will have dropped and all you have to do is snip off the tails.

Wind

(Sorry, but these jokes blow!)

Why aren't there "large-craft advisories?" How come they only care about small craft?

What a wind! Today a golfer got a hole in none!

What a wind! I saw a golfer playing golf with an anchor tied around his neck.

What a wind! Driving here I got 400 miles to the gallon.

The flight was really bumpy. When the flight attendant said, "Your lunch will be coming up shortly," she didn't know how right she'd be.

What a wind! At a barbecue today the host tossed a salad and it landed five miles away.

The wind died down for a couple of seconds . . . and everyone fell down!

The way things were today, you went anyway the wind blew you.

Because of today's wind, people were wearing Velcro on the soles of their shoes.

What a wind! In NYC it was the first time they've ever seen a traffic light go through a taxicab!

A kite is one more thing that gets where it is by pull and a lot of wind.

Clear air turbulence: Causes airline passengers not to know where their next meal is coming from!

After a stressful flight, a passenger asked the flight attendant, "Could you tell the pilot to keep that seat belt light off. Every time he turns it on it gets bumpy."

These winds are good for driving. This morning I drove 36 miles in neutral!

During a cross-country flight, the plane bounced around in the stormy sky. Flight attendants weren't even allowed to serve soft drinks. Finally as the flight descended toward the airport and below the clouds, the turbulence subsided. One passenger remarked, "I've never been so happy to be under the weather."

Winds were gusting to 30 mph. . . . That's 210 in dog mph.

It was so windy the balls hit by the Mets actually made it into the infield!

You know the winds are strong when you see a herd of cows flying south for the winter.

There used to be two windmills on this hill, but there was only enough wind for one.

The wind finally died down and people were actually taking the rocks out of their pockets.

Was it windy? No, I always have trouble keeping my eyebrows on.

Wind: A condition of the atmosphere that fluctuates so rapidly that attempts to measure it accurately have been in vane.

A perfectly calm day will turn gusty the moment you drop a $20 bill.

Becalmed: A sailing term describing the wind force and the simultaneous consumption of the last cold beer.

The wind was something. . . . I woke up this morning and I had all new lawn furniture.

Calm: A nil wind.

Jet stream: A narrow band of strong high-altitude winds that allows you to fly much faster to a place that's 500 miles from your luggage.

Mean wind: One that causes your kite to land in a tree.

Trade winds: What the USA does with Canada and Mexico.

Trade winds: They blow your old car away and leave another one in its place.

Weather vane: A Windicator.

Winter

(Jokes for when you want to do
a snow job on someone!)

I'm going south for the winter. Actually, in this wind
some parts of my body are headed there already.

Windshield scraper: A device that keeps falling out of
the glove compartment in summer, gets lost under the
seat in winter, and breaks in half the first time you use it.

Did you hear about the mathematician who turned off
his heat because he wanted to be cold and calculating?

Let me figure this out. . . . I'm saving 20% with storm
windows, 30% with insulation, 15% with weather stripping and caulking? At this rate, by spring the power
company will be sending me money!

The heating companies are planning a rate freeze this winter. You pay their rate or you freeze.

If anyone tells me our winters are getting warmer, I'll hit him with the biggest thing I can find . . . my heating bill!

Cryogenics: The branch of physics that deals with freezing bodies. It is commonly practiced by landlords and homeowners in winter.

I just received a message from the Surgeon General warning about heart failure . . . and it was on my last heating bill.

Here's a notice to all homeowners: Numbness and shock may result from exposure to your bills for heating.

With the price of heating the only place I'm hot is under the collar!

We've finally reached the balance of nature for the winter . . . it now costs as much to heat your house as it does to go to Florida for the winter.

Winter is Mother Nature's revenge for cheating on your golf score all summer.

It's the season for giving. Do you want the flu?

The flu season is going to be worse than I thought. . . . I saw a squirrel burying cough drops.

It's flu season again. A study of flu sufferers by university scientists had to be postponed when several volunteers called in well.

Flu season: The hoarse and buggy days.

It's the season you have to worry about the person who has everything . . . and they're breathing on you.

It's the time of year when the person who makes it to work through a blizzard spends the rest of the day taking calls from those who didn't.

February is to winter what Wednesday is to the workweek.

Let's hope that what happens in February stays in February.

I don't know about you, but I'm Februweary!

About the only good thing you can say about the month of February is that it only hurts for a little while.

Will February march? No but April may!

What can you say about February? It's a discount January.

I could never understand leap year. Who'd want one more day of February? Why can't we have a July 32?

February is the shortest month of the year . . . so how comes it feels like the longest?

Now you know why they made February the shortest month!

One of the worst things about February is that you can't get rid of it until March.

I love all the seasons! I love spring and fall . . . and in winter I love summer, and in summer I love winter!

The potholes in town are so big they have their own small-craft advisories!

It's always a challenge calling the highway department to complain about a pothole. They asked, "How deep is it?" I said, "How would I know?" They said, "Can't you look down?" I said, "No! I'm afraid of heights!"

I don't like to say this but potholes are getting to be a real pain in the asphalt.

Pothole repair budget: The balance of pavements.

The highway department announced that they are going to fill in 85,000 potholes. Well, at least that takes care of my street!

It always helps to remember the first rule of driving: Potholes always have the right of way.

"It's wintertime and the livin' is tricky." That's the title of the least successful song written by George Gershwin.

I finally figured out the pattern to our weather: It's warm in summer and cold in winter . . . but not necessarily.

This weather causes soreness and irritation. . . . You're sore and irritated because you didn't go to Florida!

I don't mind summer, but winter leaves me cold.

I love winter . . . you can stay inside without feeling guilty.

It's the beginning of winter . . . for meteorologists it's known as "the busy season."

I saw this sign in front of a sporting goods store, "Now is the discount of our winter tents."

In winter not only do I have to listen to people complain about the weather but I can see their breath while they're doing it.

Global warming is too slow ... I'm going to Florida!

It's that time of year when I'd feel much better if it wasn't that time of year.

It's that time of year when the car you can't start in your driveway won't stop in the street.

Winter is when fathers shovel paths to the street so their kids can get out to make money shoveling paths to the street.

Winter is when you pay a kid to shovel out your driveway so you can go to the health club to work out.

Winter is when your car won't start running and your nose won't stop.

You can't say, "Old man winter" anymore. To be politically correct, it's now "thermally challenged mature individual."

Winter is when it's too cold to do all the chores around the house it was too hot to do last summer.

I'm tired of winter . . . I can't wait for my first mosquito bite.

This forecast is as honest and accurate as the day is long . . . but keep in mind that today is one of the shortest days of the year!

I just got my car ready for winter. . . . I caulked the windows, installed a bigger muffler, and put hot chocolate in the radiator.

Every year I winterize my car. Why can't they develop something to winterize me?

Some people call it "the winter season" . . . I call it "the hangin' around waitin' for spring season."

Tomorrow is the first day of winter . . . everything up to now has been a rehearsal.

I'm promising 80° . . . how about two days with 40°?

Everyone wants to see snow. . . . I say to these people, "Why don't you just take the cable off the back of your TV and leave me alone!"

I sprained my ankle the first time I was on skis. And I was still in the sporting goods store!

Skier: A person with a two-track mind.

There's a new movie about skiing . . . everyone in the cast is in the cast.

Snow skiing? I tried that last year and couldn't get the boat up the hill.

First time I went skiing I broke a leg . . . thank goodness it wasn't mine!

Ski tow: What skiers get instead of athletes foot.

I'm not ready for downhill . . . my style is over the hill.

I would love to try snow skiing but I'm already going downhill fast!

The skier was a feminist . . . she refused to ski on man-made snow.

Skis: A pair of long, thin, flexible runners that permits a skier to slide along the snow and into debt.

Skiing is a sport where many people end up end up.

I'm starting on one of the most difficult sports there is . . . uphill skiing.

Skiing is a sport where you spend an arm and a leg to break an arm and a leg.

I did pretty well my first time on skis . . . I'm going to try it next time with snow on the ground.

Last night I dreamed I was skiing in a men's alpine event. When I woke up there were two slats missing form my Venetian blinds.

I just can't get the hang of skiing . . . for me it's an uphill battle to go downhill.

My ski instructor told me to come back when conditions are right for my style of skiing. He suggested July.

Accident policy for skiers. . . . Snow-fault insurance.

Ski resorts: An industry that prospers when business is going downhill.

My favorite winter sport is riding on a plane to Florida.

I went sledding the other day . . . if you call the seat of your pants a sled.

I was a stubborn kid. When we went sledding, I was the only kid with an inflexible flyer.

Some people can do a figure eight. I have trouble doing a figure one!

You've heard the expression "flying by the seat of your pants." That's how I ice-skate.

Ice-skating is something you can do hours on end.

Ice-skating is a winter sport preferred by people who want to hurt themselves but are afraid of heights.

Ice skates are things that take several sittings to learn how to stand on them.

This year they're going to combine swimming with luge. ... You swim some you luge some.

Skiing is a winter sport that requires white snow, green money, and Blue Cross.

This is the time of year that if you're a resident of Florida there's no such thing as a distant relative.

For you people headed south I'm predicting a winter of warm fronts ... also backs and sides!

Cold? Residents of Miami are thinking about going south for the winter.

People in Florida are not used to cold weather . . . they wear down-filled parkas in the frozen food section of the supermarket.

You know it's a bad winter when Miami weather reports start giving ski conditions.

Last year when I went to Florida I rented a car. This year I'll be prepared . . . I'm renting a sled.

In winter I just want to stay inside . . . the state of Florida.

When it comes down to it, more people depend on solar energy for snow removal than any other method.

Did you know that snow is the peanut butter of nature? It's smooth or crunchy, the kids love it, and it sticks to the roof of your house.

There's a good side to snow. In the summer your neighbor's lawn may be greener, but in winter your snow is just as white.

We can use the rain but did you ever try to build a rain man?

The dead of winter refers to my lack of feeling . . . in my toes and fingers.

I love this weather . . . a chill in the air, branches covered with snow, an exhilarating breeze . . . unless I'm outdoors!

I get more snow in my driveway than my neighbor . . . that's because I have bigger driveway.

You know you're growing old when you're no longer excited by the first snow of the season.

On this date, about 26 inches of snow buried the city. It paralyzed traffic and closed businesses, schools, and airports . . . and in general improved life around here in many other ways as well.

I'm tempted to sue Jack Frost for nipping at my nose.

A snowman fell in love with a snowwoman. He gave her a warm embrace and they melted in each other's arms!

What did the snowman say to the snowwoman? "Do you believe in love at frost sight?"

Two snowmen were talking and one said to the other "Do believe in life after thaw?

The snowmen are doing great . . . but remember it's a seasonal job.

I'm waiting for spring or global warming . . . whichever comes first.

I don't know about you but my favorite winter vacation is spring.

The leaves are gone, the birds are gone, and the flowers are gone. Do you suppose they're all smarter than we are?

I'd like to put in a good word for winter, but I can't use that word on TV.

You know that nude beach? They put up this sign, "CLOTHED FOR THE WINTER."

When I was a kid I didn't suffer from the cold weather as much. That's because I didn't keep hearing about the wind chill factor.

Did you know that winter undergarments for bike riders are known as long Honda wear?

Ah, the joys of winter . . . surely there must be some!

It's time to get out the ear muffs. My ears aren't cold . . . I'm just tired of hearing people complain about the weather.

It's the middle of the winter, when we're short on temperature and long on Johns.

Winter is when people pay $1,000 a day to go to places to get the heat they complained about in July.

Snow removal budget: The slush fund.

Where do bees go in winter? They hivernate.

Cold shoulder: The part of the highway that freezes first.

Do you have fear of a cold Friday the 13th? Then you're suffering from Briskadekaphobia.

Unusually warm spring weather: Juneuary.

A heating supplier was talking to an obstetrician about the record cold winter. He said, "January was my busiest month. People just stayed home and tried to keep warm." The obstetrician said, "Yep, mine will be in October . . . same reason."

Super Bowl forecast: Temperature LXIV and Wind XIV.

Bears sleep all winter and we think that we're the smart ones!

Weather stripping: An exotic dance performed out-doors . . . weather permitting.

Can you believe these slippery roads? It took me an hour to get to work . . . it normally takes two hours!

I've never seen it so slippery. I just drove 17 miles and never left my driveway.

Roads will be salted . . . I skid you not.

Driving on ice and snow is scary! I got to work on time but my stomach was an hour late.

Let's give the highway department a sanding ovation!

Ice fell on a taxi . . . talk about hailing a cab!

Your rear end can slip around when it's icy . . . but you probably shouldn't let your rear end get icy.

Winter travel advisory: "Fasten your sleet belt."

Winter travel advisory: "Your slip is showing."

Winter travel advisory: "Traffic moving snowly."

This weather could give your car the creeps.

I had to leave work early yesterday. My Zamboni was double parked.

Lots of people missed work today. They called in slick.

I expect more ice and snow . . . there'll be new skids on the block.

I expect more ice and snow . . . a wake-up crawl.

I expect more ice and snow . . . crawling all cars.

I expect more ice and snow . . . crawl waiting.

I expect more ice and snow . . . the crawl of the wild.

Driving in the ice and snow can be dangerous. I always find that the guy in front of me drives much too close.

I began the day putting in 30 minutes with an ice scraper . . . and that was just to clear my bathroom mirror.

There are two kinds of snowstorms: a major snowstorm and a minor snowstorm. In a minor storm, you can't get to work . . . in a major storm, you can't get to the bowling alley.

Flurries become snow when they stick together.

If you know the difference between sleet and freezing rain you've been watching too much weather on TV!

Drivers may be having nightmares, but skiers will be jumping to contusions.

That funny-looking snowman in your backyard could turn out to be your barbecue.

There's a snowman out there just waiting to be put together.

This morning I saw the most lifelike snowman I've ever seen. It turned out to be a guy waiting for a bus!

A young woman was talking to her friend about the snowman on her front lawn. "I know he's cold. But on the other hand, he never criticizes me or complains about my friends . . . and if he doesn't work out, he'll be gone in the spring."

Times have changed . . . a kid came to my door today and asked if I wanted to shovel *his* driveway!

Times have changed . . . I asked a kid to shovel my driveway and he wanted $20 for an estimate!

Light snow is when it falls on someone else's driveway.

The snow is pretty . . . only an inch or so. But no one ever closed a school because of prettiness.

Light snow is the same as regular snow except that it has 20% fewer calories.

If you're a snow lover, getting an inch of snow is like winning a dollar in the lottery.

People are singing "Frosty the Slushman."

We'll be having slush hour traffic.

Slush is snow with all the fun taken out of it.

A little kid was looking out the window at the falling snowflakes and asked his mother, "Mommy, will this be on again tomorrow at the same time?"

The only snow removal system I care about is spring.

The streets are in deplowable shape . . . if you get my drift.

The higher the drift, the harder it is to find someone with a snow shovel.

If salt doesn't melt the snow . . . try a little paprika or basil.

An even greater disruption of a snowstorm is caused by those who make it to work and spend the rest of the day talking about how they did it.

This is great weather for hunting to see if you can find your car.

One snowflake to another: Stick with me and pretty soon you'll get the drift.

How come snow banks are never overdrawn?

I have no objection to snow falling . . . what I object to is me falling!

Anyone who says "no two snowflakes are alike" never saw a stadium full of football fans.

The biggest problem is getting used to calling your attic window your front door.

This is the kind of morning you get up, look outside, dial your office, and call in warm.

There will be 26 feet of snow on your driveway. That's width not height. . . . I'm the only meteorologist who forecasts the width of the snow.

No two snowflakes may be alike . . . but they all taste alike!

I wouldn't mind all this ice, snow, and slush if it came in July when the weather was nice.

Remember when you were a kid and told that if you dug deep enough you'd hit China? With this snow, if you dig deep enough you'll hit your driveway!

It's the first time they ever had snow in Monaco. Their snow plow is a Porsche with a spatula on the front.

That gated community is so exclusive the snow comes in decorator colors.

Do you know what they do in Buffalo when it snows? They let it.

We have Disney Weather . . . *Snow White and the Seven Drifts.*

Isn't this snow beautiful? But beauty is only shin deep.

Snowstorm: It's a case of driveway robbery.

Snowstorm: Weather that goes from lovely to shovely.

Snowflakes are one of nature's most fragile things, but look what they can do when they stick together. It's the best example of the adage "In union there is strength!"

Snow removal complaints: When there's a lot more open mouths than open streets.

Off to school tomorrow . . . you'll hear the pitter patter of little sleet.

The abominable snowman: The guy in your neighborhood who won't shovel his walk.

Do you know that snow is a great equalizer? No matter how prestigious your neighborhood is, after a snowstorm it becomes skid row.

On a snowy night, parents read bedtime stories to their kids about the little snowplow that could.

You have to look at the good side of snow . . . it fills in all the potholes!

Just what we need around here . . . more flakes!

Did you ever wonder where the white goes when the snow melts?

Everyone always asks me if it' snowing outside . . . but how often does it snow inside?

My neighbor is into health foods. He uses salt substitute on his driveway.

Some people like the snow . . . obviously they're flaky.

We'll be getting frosted flakes for breakfast.

You know you're growing old when you never have the urge to throw a snowball.

Heavy snow is one of the few pleasures left to people who don't drive.

How does the guy who drives the snowplow get to work?

Not only are there no two snowflakes alike . . . studies have shown there are no two goose bumps alike either.

The Packers are well known to have the best defense against the blitz . . . it's called 6 feet of snow.

Times are changing. My neighbor just bought his kid an artificial snowman.

Snow can cause a heart attack. If you doubt it just ask the kid next door what he'll charge to shovel your walk.

Environmentalists are angry over the way we're messing up nature. Then along comes a snowstorm to remind us that it's a two-way street.

The father heard that middle-aged men should not shovel snow because of the risk of heart attack. He asked his teenage son, "Would you do it for me?" The son said, "Sure immediately . . ." and the father nearly had a heart attack.

Sign at a New England resort: "Come up and ski me sometime."

I'm a real advocate of solar energy especially when it comes to removing snow from my driveway.

It's been said that the snow falls upon the just and the unjust alike . . . after which, the just fall upon the snow that the unjust haven't cleared away.

The two things I look forward to in winter are the first snow . . . and the last!

Here's a helpful hint for winter: If you call for pizza enough times you won't have to shovel your walk.

This is the time of year I gaze upon the snow drifts and ponder . . . somewhere out there is my lawn furniture.

How can Northerners get excited about winter sports like ice-skating, cross-country skiing, and tobogganing? To us it's called "going to work!"

I can program my new coffee maker. It starts making coffee before I wake up. I'm looking for a snow blower that works the same way.

If you own a snow blower, here's how it works: You blow your snow onto your neighbor's driveway. Then he blows it onto his neighbor's driveway, and on and on. The next thing you know, some guy in Miami is wondering where all this snow came from.

Seven days of shoveling snow makes one weak.

It's time to go back to work. I just have to remember which pile of snow my car is buried under.

Snowplow: A mechanical device used to pile up large banks of snow at the end of your driveway as soon as you're finished shoveling it.

I had this crazy dream last night. I was in Panama and it was snowing. I was dreaming of a white isthmus.

Icicle: A stiff upper drip.

Frozen rain: The pitter patter of little sleet.

Last night's blizzard was enough to winterupt traffic.

The worst part of the snowstorm is listening to TV forecasters trying to come up with clever nicknames like Snowmageddon and Snowpocalypse.

After a winter without much snow, a grandfather was telling his grandchild, "When I was a little kid. I'd walk six miles through the snow to get to school." The child asked, "They had snow even back then?"

It's interesting the way cities deal with snowstorms. First the snowplow comes through and pushes the snow off to the side and blocks the driveways. Then homeowners shovel it out into the street again. Four hours later the snowplow comes back and pushes it off to the side again, blocking the driveways. The homeowners shovel it out into the street again. The next day the plow comes through and pushes it to the side and the homeowners shovel it into the center again. We really don't remove snow. We nag it to death.

What can I say about winter in New York City? It's a nice place to visit, but I wouldn't want to shovel there.

One winter morning, a couple was listening to the radio while having breakfast. They heard, "We're going to have 8 to 10 inches of snow today. You must park your car on the even-numbered side of the street so the plows can get through." So the woman goes out and moves her car. A week later, the announcer says, "We're expecting 10 to 12 inches of snow today. You must park your car on the odd-numbered side of the street." She goes out and moves her car. A few days later, the radio announcer says, "We're predicting 12 to 14 inches of snow. You must . . . ," as the power goes out. The woman is upset and says, "I don't know what do! Which side do I need to park so the snow plows can get through?" The man says, "Why don't you just leave the car in the garage this time?"

Remember when kids used to pick up extra cash clearing sidewalks after a snowstorm? Now if you hand them a shovel, they want to know if it comes with batteries or where can they plug it in!

Snow makes people stupid. If you're driving on icy roads and you start to slide, what do people tell you to do? Turn your wheels in the direction you're sliding. It's like saying, "If you see a tractor trailer coming head-on, floor it."

With the cost of heating your home, people are actually asking Santa to put coal in their stocking.

When I was a kid I caught my father in one of those Dad lies. "When I was a kid I used to walk to school three miles in the snow." So I said, "But dad, we live in Miami."

Avalanche: Runaway terrain.

Blizzard: The bad news brrrs.

Immobile: Snowbound in Alabama.

Warehouse: Cry of a person lost in a blizzard.

This is a tough time for meteorologists. Everyone wants to know if we'll have a white Christmas . . . Irving Berlin had to go and open his big mouth!

I just hope Santa has an instrument rating!

I just love an old-fashioned Christmas season . . . snow on the ground, a chill in the air, and chestnuts roasting in a microwave.

Because of this gloomy weather, Santa was diverted to Atlanta where he had to change reindeer.

The stores are starting to have their holiday . . . it's the winter of our discount-ent.

Sporting goods stores are having their annual sales. I saw this sign: "IT'S THE DISCOUNT OF OUR WINTER TENTS."

If it snows on Hanukkah, the highway department announced they would use kosher salt on the roads.

How come there are no snowwomen?

Christmas forecast: Freezing's greetings.

Fleece Navidad.

Christmas rain: H_2 Ho Ho Ho.

The Christmas shopping season is over. You're now in the owe zone layer.

Have you noticed that all the presents these days are electronic? Years ago the biggest problem you had on a cold Christmas morning was getting your car started. Now it's getting the toys started.

Christmas forecast: Toynadoes.

Christmas this year comes on Toysday.

No doubt about it. 2013 has been an odd year.

There'll be two inches of water on New Year's Eve . . . most of it will be sparkling.

For my New Year's resolution, I decided to be more decisive when I make my forecasts. Do you think that's a good idea?

I'll make a prediction. The rotten weather of this year will end on December 31 and become the rotten weather of next year.

My prediction is . . . sometime during this year the weather will be unusual.

If the groundhog can tell when winter's over . . . he can have my job.

It's Groundhog Day . . . but enough about the school lunch menu.

The groundhog on Noah's Ark predicted partly cloudy!

The groundhog saw his shadow and said it made him look fat.

The groundhog is nothing more than a woodchuck with a publicity agent.

Emergency: What the groundhog will do on February 2.

The groundhog is an animal with a woodchuck's body and an old wives' tale.

The groundhog saw his shadow, which means six more weeks of basketball and hockey.

The prediction made by the groundhog is known as the "furcast."

Groundhog Day is when the media gather to show they have too much time on their hands.

How can you rely on someone who's afraid of his own shadow?

The Las Vegas groundhog not only tells you if you have six more weeks of winter but it gives you the odds and the temperature spread.

Who do you trust more, a rodent or a bona fide meteorologist?

Now that Groundhog Day is over, Phil can go back to his regular job as commodities broker for a Wall Street firm.

If Punxatawney Phil comes out and sees his shadow, we'll have six more weeks of winter. However, if he comes out and sees Punxatawney Phyllis we'll have six more groundhogs!

After predicting six more weeks of winter, he also predicted a tax increase, a bear market, and a 5 percent increase in the unemployment rate.

Today is Groundhog Day . . . to see or not to see . . . only the shadow knows!

February 29 is a day created by our government because they couldn't cram all the rotten weather into just 28 days.

What did the snowman say to the beautiful snow-woman? Do you believe in love at frost site?

Love is in the air . . . as if the air doesn't have enough problems!

For the Chinese New Year we'll have Chinese weather . . . dim sun.

Abraham Lincoln walked to school every day, 12 miles there and 12 miles back, no matter whether there was rain, sleet, or snow. No school buses. No hot lunches. What a lousy PTA that school must have had!

Seeing people's breath when they complain about the weather is one of a meteorologist's favorite things.

If it wasn't for goose bumps I wouldn't have any physique at all!

I love this weather ... I can see my breath and that's reassuring for a person my age!

Better bundle up kids—it'll make your mother feel warm all over.

Mothers say to their kids, "I'm cold, so put on your sweater."

My nose is froze' and so's my toes.

One worker sneaked over to turn up the thermostat ... but he was caught blue handed.

I never thought I'd see the day when more employees are watching the thermostat than the clock.

I wonder if management can make this room-temperature?

Just think the heat we pay so much for, was free last summer.

You can always tell how cold it is by how many animals end up in bed with you.

"Cold? If the thermometer had been an inch longer we'd all have frozen to death." (Mark Twain)

I don't mind exchanging ballet, technology, and art with the Russians. But do we have to exchange weather?

What did I tell you? The cooler weather you wanted back in August finally made it.

How does it feel to be one of God's frozen people?

Not only has the iceman cometh . . . I think he's decided to stayeth.

You'll have to excuse me . . . the only way I can describe this weather is with a four-letter word. . . . *Brrr*!

Where's the greenhouse effect when you really need it?

To keep your feet warm, the best thing to do is dip the affected area in warm water . . . preferably in Florida!

If this weather is getting to you, it might help to think that in about a couple of months you'll be complaining about your AC bills.

It's so cold . . . this would be a good time to start a heated argument.

Remember this if you can't stand the heat . . . you're going to love this forecast.

I don't know why people are complaining about the cold. Today it was 75° . . . 65° inside and 10° outside.

Cold front: What you get when you stand with your back to the fireplace.

They say "everyone talks about the weather." But that's not easy when your teeth are chattering!

My forecast will have a few well-frozen words.

You know it's cold when Starbucks is serving coffee-on-a-stick.

The temperature dropped so fast the weather service reported skid marks on its thermometer.

There won't be any forecast tonight . . . Jack Frost was nipping at my notes.

It was so cold . . . you could freeze an egg on the sidewalk.

It was so cold . . . I went outside for a bite of coffee.

It was so cold . . . my frosted flakes had real frost on them.

It was so cold . . . I sent my friend a birthday card and my tongue is still on the envelope.

It was so cold . . . the Giants went into a huddle and didn't want to come out.

It was so cold . . . after the game the coach was hit by a block of Gatorade.

It was so cold . . . I saw a snowman building a snowwoman.

It was so cold . . . they performed Hamlet on ice.

It was so cold . . . people were getting on planes to Florida and not even asking if there's a lower fare.

It was so cold . . . the only thing that kept me warm was my heartburn.

It was so cold . . . Superman froze his S off.

It was so cold . . . people in Texas were wearing 10-gallon earmuffs.

It was so cold . . . last night I put on my coat to take out the garbage and it didn't want to go.

It was so cold I had a piece of ice cream stuck in my teeth.

Do you know how Alaskans dress? As quickly as possible!

We're right on schedule . . . winter's half over and I'm half frozen.

Does shivering count as exercise?

The only thing holding me together is static cling.

It's a good thing the power company doesn't charge us for static electricity.

Static electricity: Science friction.

The National Weather Service has issued a static cling advisory.

The question is: Is static electricity better than no electricity at all?

Did you ever wonder which came first, static electricity or door knobs?

I asked my librarian for a book on static electricity. She asked me if I wanted friction or nonfriction?

It hasn't been this cold since they invented the wind chill factor.

Right now it's a comfortable 35°. Unless you go outside, and then it's cold and nasty.

It's cold. Read my breath.

You're not a kid anymore when you're obsessed by your thermostat.

I can see my breath. It's either cold out there or I need to get a better mouthwash.

When the kids go out to play, remember they can only go as far as the extension cord will allow.

When I turned on my shower today I got sleet.

This frigid weather is coming off a cold front between the Democrats and the Republicans.

Cold? I'm wearing gloves, a scarf, and thermal underwear. I won't even go near draft beer!

Winter theme song: "Freeze a Jolly Good Fellow."

The Dow jumped 60 points. If you factor in the winds, though, it felt like 40 points.

It's really cold. If you decided to wear your gloves and scarf you've made a wise choice. You've made a wiser choice if you decided to wear the rest of your clothes too.

Electric blanket: A device whose owner is asleep at the switch.

I'm predicting more days of bitter cold and . . .
. . . shivery is not dead.
. . . it's enough to chill a mockingbird.
. . . with temperatures in the tingle digits.
. . . we'll be doing dances with wools.
. . . we'll be having shivers regal.

Camping out this week? The cold will be in tents.

If you see me standing in front of the refrigerator with the door open, I'm just trying to warm up.

I hate a cold front. I also hate a cold rear.

Old snowmen never die . . . they just melt away.

The secret to surviving in very cold weather is to put on so many layers that you can't get through the door to go outside.

Never get emotionally involved with a snowman.

Now he's de-frosty the snowman.

Meteorologists

(Didn't the Editor say in the Introduction
there are no "Meteorologist Jokes"?)

Where do meteorologists go for happy hour? The Iso-
Bar! (Attributed to Adam Arnold)

Where do meteorologists go for a very quick, short
drink? The Millibar! (Attributed to Adam Arnold)

Why do meteorologists seek help from psychol-
ogists? Sometimes they get depressions! (Dennis
Negron-Rivera)

Did you hear about the meteorology grad student
who was killed drinking milk last week? NO! How in
the world did that happen? Cow fell on him! (George
Greenly)

What happened to the meteorologist after a dry forecast busted? He was all washed up! (Alek Krautmann)

Meteorologist: This guy had such a great deal on barometers, I bought seven of them! Friend: SEVEN barometers? Why on earth did you buy so many? Meteorologist: He was a high-pressure salesman! (David Keller)

Do you know how to tell an introverted meteorologist from an extroverted meteorologist? When an extroverted meteorologist talks, he looks at YOUR shoes. (David Salmon)

The Last Joke

(You might just need a meteorologist
to explain this one!)

Two guys sit down at a bar. The first guy says, "I'll have
an H_2O." The second guy says, "Hey that sounds good!"
And so he says to the bartender, "I'll have an H_2O too."
... And he died! (As told by Bill Hooke)